MATERIALS SCIENCE AND TECHNOLOGIES

FERROFLUIDS

MATERIALS SCIENCE AND TECHNOLOGIES

Additional books in this series can be found on Nova's website
under the Series tab.

Additional e-books in this series can be found on Nova's website
under the e-book tab.

MATERIALS SCIENCE AND TECHNOLOGIES

FERROFLUIDS

FRANCO F. ORSUCCI
AND
NICOLETTA SALA
EDITORS

Copyright © 2013 by Nova Science Publishers, Inc.

All rights reserved. No part of this book may be reproduced, stored in a retrieval system or transmitted in any form or by any means: electronic, electrostatic, magnetic, tape, mechanical photocopying, recording or otherwise without the written permission of the Publisher.

For permission to use material from this book please contact us:
Telephone 631-231-7269; Fax 631-231-8175
Web Site: http://www.novapublishers.com

NOTICE TO THE READER

The Publisher has taken reasonable care in the preparation of this book, but makes no expressed or implied warranty of any kind and assumes no responsibility for any errors or omissions. No liability is assumed for incidental or consequential damages in connection with or arising out of information contained in this book. The Publisher shall not be liable for any special, consequential, or exemplary damages resulting, in whole or in part, from the readers' use of, or reliance upon, this material. Any parts of this book based on government reports are so indicated and copyright is claimed for those parts to the extent applicable to compilations of such works.

Independent verification should be sought for any data, advice or recommendations contained in this book. In addition, no responsibility is assumed by the publisher for any injury and/or damage to persons or property arising from any methods, products, instructions, ideas or otherwise contained in this publication.

This publication is designed to provide accurate and authoritative information with regard to the subject matter covered herein. It is sold with the clear understanding that the Publisher is not engaged in rendering legal or any other professional services. If legal or any other expert assistance is required, the services of a competent person should be sought. FROM A DECLARATION OF PARTICIPANTS JOINTLY ADOPTED BY A COMMITTEE OF THE AMERICAN BAR ASSOCIATION AND A COMMITTEE OF PUBLISHERS.

Additional color graphics may be available in the e-book version of this book.

Library of Congress Cataloging-in-Publication Data

ISBN: 978-1-62808-410-8

Published by Nova Science Publishers, Inc. † New York

CONTENTS

Preface		**vii**
Chapter 1	Introduction to Ferrofluids *Swapna Nair and M. R. Anantharaman*	**1**
Chapter 2	Magnetism: General Ideas *Swapna Nair and M. R. Anantharaman*	**13**
Chapter 3	Ferrofluids - Synthesis Techniques *Swapna Nair and M. R. Anantharaman*	**29**
Chapter 4	Optical Properties of Ferrofluids: Grain Size - The Determining Factor *Swapna Nair and M. R. Anantharaman*	**39**
Chapter 5	Non Linear Optical Properties of Ferrofluids Investigated by Z Scan Technique *Swapna Nair, Jinto Thomas, Suchand Sandeep,* *M. R. Anantharaman and Reji Philip*	**53**
Chapter 6	Superparamagnetic Iron Oxide Nanoparticles Based Aqueous Ferrofluids for Biomedical Applications *Mary A. P. Reena and M. R. Anantharaman*	**77**
Chapter 7	Industrial Applications of Ferrofluids *Swapna Nair*	**97**
Chapter 8	Microwave Properties of Ferrofluids *K. Sudheendran and Swapna Nair*	**107**
Chapter 9	Ferrofluids Based 1D Magnetic Nanostructures *T. N. Narayanan*	**119**
Chapter 10	Proposed Multiferroic/Magneto Electric Fluids: Prospects, Techniques and Probable Candidates *Swapna Nair, Radheshyam Rai and K. Sudheendran*	**139**
Index		**149**

PREFACE

Four technologies are going to rule the world in the 21st century of which the first three are biotechnology, technology of photonics and genetic engineering which had shocked the world with their new findings in the late 90's itself and showing their potential in day to day human life. Nanotechnology is the fourth one which has the additional advantage of being interdisciplinary to the first three. Nanotechnology deals with the phenomena and structures that can only occur at the nanometer scale which is the scale to represent single atoms/molecules. Hence it can be defined as the synthesis and engineering of materials near the molecular level. Nanometric scale is used to specify materials at a size range 1 to 100nm where unique phenomena is found to be exhibited which enables novel applications. The related term nanoscience which is an extension of 'materials science' is used to describe the interdisciplinary fields of science devoted to the study of nanoscale phenomena employed in nanotechnology consisting of processing, separation, consolidation and deformation of materials and induced property change at the molecular level.

The first mention of some of the distinguishing concepts in nanotechnology was by Richard Feynman in a lecture at an APS meeting in 1959 which started with a sentence "there is plenty of room at the bottom". He described a method by which one can manipulate individual atoms and molecules in which he also talked about the role of surface in determining the bulk behavior.

Magnetic nanomaterials have become the topic of interest nowadays due to its unique application potential in versatile fields including high density storage media, magneto-optical display devices etc, due to their very high surface area. Diluted magnetic semiconductors are another promising area which has great application potential. Nanograined magnetic materials show many peculiar properties like single domain nature, superparamagnetism, spin polarized tunneling etc, which can open a new era in spin modulated electronics (spintronics). Also these ultra fine magnetic materials have potential bio-medical application including in targeted drug delivery, hyperthermia etc due to their additional advantage of being functionalized and targeted by the application of external magnetic fields.

Also when the grain size of these nanostructures approaches Bohr radius of the materials, their band structure gets modified to a great extend giving their optical properties remarkably different from their bulk cousins. Metals if synthesized in the ultrafine regime are no more metals! They are semiconductors. Magnetic oxides which are semiconductors optically whose bandgap varies from 2.1 eV to 2.2 eV in the bulk, can be modified to a great extent if they could be synthesized with in the Bohr radius limit due to quantum confinement effects. Thus tailoring the grain size can ensure fine tuning of the magnetic and optical properties which can enhance their application potential.

The important properties that make the nanoscaled magnetic materials important from the fundamental point of view is the remarkable modifications in the structural, morphological, chemical, optical and magnetic properties at the nanometric regime when compared to their bulk counter parts. Magnetic nanomaterials, especially the magnetite derivatives are thus ideal templates to study the optics, magneto-optics and magnetism at the nanolevel. However, a liquid, which possess the qualities of magnetic nanomaterials with its added advantage of their flow control in presence of an applied magnetic field thus becomes quite important from the application point of view.

Ferrofluids are stable colloidal suspension of nanosized magnetic materials in a suitable basefluid. They can be synthesized with or with out suitable surface coatings. The properties are greatly influenced by the grain size, carrier fluid chosen, and composition of the magnetic material and concentration of them in the fluid. Their flow and properties can be tuned with the application of an external magnetic field. The application potential of ferrofluids is enormous. They are important templates from the fundamental physics point of view to study the magnetism of non interacting nanodomains. Also by tuning the concentration of this system, the effect of interaction also could be studied. They are also ideal templates for studying the quantum confinement effects in nanomagnetic materials. Hence, optically they are ideal system as the grain size approaches the Bohr radius limits, high quantum confinement and induced blue shift can make them optically more transparent and this confined excitons can assist in non linear absorptions. As a random media, this can offer induced scattering which can make them good optical limiters owing to their high shelf life and stability against agglomeration.

Thus the book contains 10 chapters. In this book, the former sections explains the general properties exhibited by ferrofluids, while the latter section deals with the application of ferrofluids like its application in industry, biology and medicine, and other areas. Magnetic, optical, non linear optical and microwave properties of ferrofluids are also explained in this book, with special thrust on the integration of these properties into device formaulation. Also a futuristic proposal of the synthesis of a multiferroic or magneto-electric fluid is put forwarded towards the end.

Chapter 1 gives an introduction to ferrofluids with necessary details and Physics governing their properties. General properties and applications are given in brief in this chapter.

General theory of magnetism is provided in detail in *Chapter 2*. The phenomena happening in nanosized magnetic particles are highlighted in the latter part of this chapter.

Chapter 3 deals with the general synthesis techniques that are adopted for the synthesis of ferrofluids. A detailed literature survey is provided to include a lot of possible synthesis methods. A god schematic diagram for the most adopted chemical technique is also given.

The optical properties of ferrofluids in the linear regime are detailed in *chapter 4*. The role of confinement on the optical properties of ferrofluids is explained in this chapter. A detailed theoretical description of quantum size effects is also provided so as to help the readers in understanding the quantum confinement effects in ferrofluids easily.

Chapter 5 describes the non linear optical properties of ferrofluids and explains their application in optical limiting devices. Necessary experimental details and theoretical concepts for better understanding of the non linear optical properties is provided in this chapter.

Chapter 6 deals with the biomedical application of ferrofluids. Ferrofluids are widely employed in biology and medicine and some of the important application of ferrofluids are in cancer therapy like targeted drug delivery, hyperthermia etc. Some experimental results conducted in authors group is provided in detail in this chapter.

Industrial applications of ferrofluids are enormous. Right from their application in optical and magneto topical display devices, art, application in sealing, as lubricants and coolants, etc. are only some among them. *chapter 7* details such application of ferrofluids.

Microwaves (300 M Hz – 300 G Hz) have become very important for today's human life and which are extensively used in today's civilian and military communication systems as well as in domestic and industrial appliances. However, microwaves are detrimental to human life if subjected to sufficient intensity and longer exposure. Hence the microwave absorbing devices becomes quite important in today's life. *Chapter 8* gives the microwave absorbing properties of ferrofluids. Authors included some of the unpublished works from our group in this chapter.

Chapter 9 provides a detailed summary of the research work done so far in ferrofluid based 1-D structures and their application in different fields.

Multiferroics can be primarily defined as single phase materials which simultaneously possess two or more primary ferroic viz. ferroelectric, ferromagnetic, ferroelastic and ferrotoroidic properties. Of this, magneto-electric materials form a group, which possess vast and versatile application potential. However, a fluid with multiferroic/magneto-electric properties is seldom /not investigated so far. *Chapter 10* provides an introduction to such proposed fluids and some probable candidates. Some techniques are introduced in this chapter for their synthesis. A brief theory of multiferroic and magneto electric materials is provided for the authors.

Thus this book serves as a good guide for the beginners working in the area of ferrofluids.

In: Ferrofluids
Editors: Franco F. Orsucci and Nicoletta Sala

ISBN: 978-1-62808-410-8
© 2013 Nova Science Publishers, Inc.

Chapter 1

INTRODUCTION TO FERROFLUIDS

Swapna Nair[1,] and M. R. Anantharaman[2]*
[1]Departmento de Engenharia Cerâmica e do Vidro and CICECO,
Universidade de Aveiro, Aveiro, Portugal
[2]Department of Physics,
Cochin University of Science and Technology Kochi, India

ABSTRACT

This chapter gives a general introduction to ferrofluids. Ferrofluids Although magnetism and magnetic materials have been hot research topic as early as 1900, the realization of a fluid exhibiting magnetic properties was only a dream of researchers till 1960. Several researches were done in this regard after that, and mainly top down methods (reducing the particle size by continuous grinding process in a roller mill) were dominated in those days. However, those were really time consuming methods. The enthusiasm in this area was grown in multitudes with the invention of cost effective and faster methods for their synthesis, like chemical co precipitation. This chapter begins with the need of such a fluid, and the early researches that were carried out. This chapter covers the general properties, synthesis methods and application of ferrofluids. Basic physical laws governing the stability criteria and flow is also dealt in this chapter.

INTRODUCTION

A homogeneous fluid with magnetic characteristics has been a dream of researchers and scientists for quite long which remained unfulfilled as efforts for producing a liquid out of a solid by melting the metallic magnets became impractical because all the magnetic materials have a property of loosing their magnetic properties if heated above a certain temperature called the Curie temperature. These particles have melting point much greater than the Curie temperature, twice as that of the value of its Curie temperature. This makes the process of producing magnetic fluids by melting, a non-viable one.

---

[*] E-mail: swapna.s.nair@gmail.com.

The dream of producing a liquid that possess strong magnetic properties was not realized until the early 1960s, when Stephen Papell of National Aeronautics and Space Administration (NASA) developed a colloidal system. Papell's fluid consists of finely divided particles of magnetite suspended in kerosene. To keep the particles from clumping together, Papell added oleic acid, an organic substance that served as surfactant or dispersing agent [1].

This kerosene-based fluid had a high evaporation rate and was not suitable for industrial applications. After a few years of NASA funded magnetic colloid research at AVCO, Ferrofluid Corporation was founded by Rosensweig and Moskowitz, to commercialise this technology. Magnetic fluids are of great interest, since they possess the properties of a fluid and act as a ferromagnetic material. [1-10]

COMPONENTS OF A MAGNETIC FLUID

The unique combination of fluidity and the capability of interacting with a magnetic field is achieved in magnetic fluids because of their composition. Three components are required to synthesize a magnetic fluid namely, a liquid base (or in other words, a carrier liquid), single domain magnetic particles of a colloidal size and a stabilizer to prevent colloid particles from aggregating. Each of these components must satisfy certain requirements [2, 3].

(a) Base Fluid

A carrier liquid is chosen to conform to its field of application. Thus for lubrication and sealing systems, mineral oils and silicon organic bases are used. For medical applications, water is used as base fluid. Liquid bases need to be of low evaporation, non toxic, resistant to corrosive media, insoluble in specified media and so on [2, 3].

(b) Single Domain Magnetic Particles

Ferromagnetic particles in colloidal dispersion make the fluid act like a ferromagnetic material. They may be cobalt, iron, nickel or one of their magnetic compounds or alloys. The most usual material is magnetite. Typically magnetic fluids contain 10^{20} particles per litre [2, 3]. The size of the particles must be sufficiently small, for preventing agglomeration and precipitation. The thermal motion of the particles ensures the stability of magnetic fluid and this thermal motion increases with decreasing particle size. At the same time the particles must not be too small, since at sizes less than 1-2 nm, their magnetic properties disappear [2, 3, 11].

(c) Surfactant (Stabilizer)

The surfactant must prevent particles from aggregating. To this end long chain molecules are used with functional groups OOH, H_2OH, H_2NH_2 and so on. A stabilizer is chosen so that its molecules interact with magnetic particles, via bonds of functional group, to form a tightly bonded monomolecular layer around the particles [2-4].

What is a Ferrofluid?

Ferrofluids are stable colloidal suspension of single domain magnetic particles in a base fluid, which is magnetically passive. The number density of the particle in a typical ferrofluid is of the order of 10^{23} particles per m^3. The particles are so small in nature and the size of the particles is of the order of 100 Å [1-4].

To synthesize ferrofluids, at least two components, mono domain magnetic particles and a suitable carrier liquid are required. Since randomising Brownian energy may not be enough to counteract attraction owing to van-der-Waal and dipole- dipole forces, aggregation and sedimentation are prevented by providing suitable repulsive forces either by Coulomb repulsion or by steric repulsion. In the former case particles are either positively charged or negatively charged and the fluid is called 'ionic ferrofluid', while in the later case each particle is coated with an appropriate surfactant and resulting ferrofluid is classified as 'surfacted' ferrofluid. The ionic fluid requires a polar medium like water as base fluid; the surfacted fluid can use any carrier liquid like oil, water and hydrocarbons [12].

The choice of the carrier liquid depends on the application. For a surfacted ferrofluid, selection of surfactant is crucial for its stability. A surfactant molecule consists of a polar head and a tail of hydrocarbon chain. An example is oleic acid

$$CH_3(CH_2)_7 CH-CH(CH_2)_7 C \begin{smallmatrix} OH \\ O \end{smallmatrix}$$

The anchor polar group is adsorbed on the particle while the chain performs thermal movement in the carrier. When a second particle with a similar chain approaches closely, the movement of the chains is restricted and results in steric repulsion.

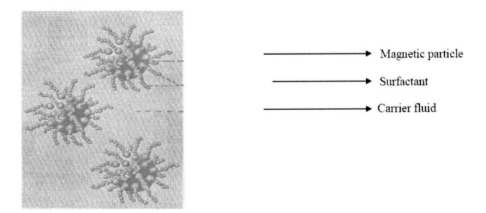

Figure 1. Magnetic particles suspended in a carrier liquid.

Rosenweig gives a simplified estimate for the entropic effect for the short chains

$$E_S = \frac{2}{3}\pi N k_B T \left[\delta - \frac{x}{2} \right]^2 \frac{\left(1.5D + 2\delta + \frac{x}{2}\right)}{\delta}$$

(1.1)

where N, is number of molecules per unit area, δ, is the thickness of the stabilized layer, x is the distance between two surfaces, and D is the diameter of the particle. The equation is useful to calculate δ which gives a reasonable barrier of $25k_B T$ to prevent agglomeration [2, 3].

PHYSICS OF FERROFLUIDS

Many of the properties of the ferrofluids can only be well understood by studying the physical laws governing the behaviour of these special fluids. So a brief discussion of major physical laws governing their special properties becomes relevant here. This section mainly consists of three parts, namely, Modified Bernoulli's equation, Kinematics of ferrofluids and Brownian motion of the particles in a ferrofluid.

Stability Criteria for the Magnetite Based Ferrofluids

Stability against settling of the suspended particles is achieved if the ratio of thermal energy to magnetic energy is greater than 1, ie,

$$\frac{k_B T}{\mu_0 MHV} \geq 1$$

(1.2)

where M is the magnetisation, H is the applied magnetic field and V the volume of the particles. Assuming the particles to be spherical, the diameter of individual particles is given by

$$d \leq \left(\frac{6k_B T}{\pi \mu_0 MH}\right)^{1/3}$$

(1.3)

so that with a permanent magnet having a magnetic field of 8×10^4 A/m, we can separate the aggregates to have an average particle diameter around 80 Å.

Stability against individual particle agglomeration is achieved if the thermal energy becomes at least equal to the dipole-dipole energy:

$$\frac{\mu_0 M^2 V}{24} \leq k_B T$$

(1.4)

and for magnetite particles the magnetization M= 4.46 x 10^5 A/m, giving the maximum size limit for an agglomeration less ferrofluid as 100 Å.

MODIFIED BERNOULLI'S EQUATION

The equation presented by the Swiss mathematician Daniel Bernoulli in his Hydrodynamica of 1738, is one of the most useful relations in ordinary fluid mechanics. This equation relates the pressure, the velocity and the elevation of a fluid in a gravitational field. Bernoulli showed that the sum of the three forms of energy; pressure energy, kinetic energy and gravitational energy inherent in the flow remains constant, provided, the effects of friction are negligible [13].

The Bernoulli's equation is

$$P + \frac{1}{2}\rho v^2 + \rho g h = const. \tag{1.5}$$

Hydrodynamically magnetic fluids follow the Bernoulli's equation, modified by adding a term, which takes the magnetic properties into consideration. The pressure energy, potential energy and magnetic energy are constant along the streamline flow. It is expressed by the modified Bernoulli's equation,

$$P + \frac{1}{2}\rho v^2 + \rho g h - \mu_0 \int_0^H M dH = const. \tag{1.6}$$

where P is the pressure energy, ρ is the mass per unit volume, μ_0 is the permeability of free space, M is the magnetization and H is the magnetizing field. From this equation it is clear that in the absence of magnetic field, the magnetic fluid acts like other liquids, but in magnetic field an additional force appears and affects the fluid. [2, 13].

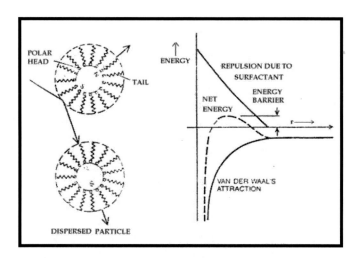

Figure 2. Various interaction energies those are significant in the prevention of particle agglomeration.

Kinematics of Ferrofluids

The kinetic theory of matter suggests that sufficiently fine particles of matter can remain suspended indefinitely in a liquid even though the particle density is much higher than that of the liquid. This is due to the continuous collisions of the particles with molecules of liquid, which are in thermal motion. When many particles are present in the same volume of the liquid, agglomeration may take place, creating a heavy particle, since the ratio of the thermal energy (kT) to the gravitational energy (ρgh) is reduced.

The forces of attraction that affects the particles in a ferrofluid are magnetic force and Van der Waal's force. Due to the extremely small size of the particles, the magnetic force is very weak. But Van der Waal's force of attraction arises between the dipoles, which are created by the fluctuations in electronic structure. According to Fritz London, the energy needed to overcome the Van der Waal's force is inversely proportional to the sixth power of the centre-to-centre separation of the particles. Hence to overcome this force, the particle must be kept well apart. In preparing ferrofluids the necessary separation can be achieved by coating each particle with a molecular film that acts as an elastic cushion. [1-3].

Brownian Motion of Particles

The problem of agglomeration of the particles in a ferrofluid due to Van der Waal's force is overcome by the coating of the surfactant layer on each particle. But to avoid the gravitational settling of the particles, the gravitational energy must be exactly balance the viscous force of the base fluid. For that, there is a maximum limit for the size of the suspended particles. Then only there is a stability of the fluid by means of the vigorous non-uniform zigzag motion of the particles inside the base fluid. This is called the 'Brownian motion' of the particles [1-3]

For balancing the viscous force, which acts upward must at least be equal to the gravitational force.

Viscous force = gravitational force

$$6\pi\eta av = mg \qquad (1.7)$$

where mg is the weight of the particle, which acts in downward direction.

$$6\pi\eta av = \frac{4}{3}\,\pi a^3 \rho g \qquad (1.8)$$

Here ρ is the density of the particle, η is the co-efficient of viscosity, 'a' is the radius of the particle. Substituting the typical values, assuming the number density is of the order of 10^{23} per m^3, we get the particle size maximum of 100Å [2, 3].

Physical Properties of Ferrofluids

- The size of the particles in the ferrofluid is of the order of 100Å. They are single domain particles and exhibits superparamagnetism.
- The number density of the ferrofluids is of the order of 10^{23} per m^3. It is comparable with the number of molecules in air at STP (ie; 2.7×10^{23}/m^3)
- Physical properties of the ferrofluids like specific gravity, viscosity, magnetization, dielectric constant etc. are modified with the application of a magnetic field.
- Depending on the choice of the surfactant and base fluid, density of ferrofluid varies with the density of water to twice its value.
- Magnetic fluids have zero remanance and coercivity.
- They spike under the influence of a magnetic field.
- Cluster Formation:
  In the absence of magnetic field, the particles are well dispersed. When a weak field is applied, particle tends to form tiny clusters elongated parallel to the field. These are called micro clusters. Further increase in the magnetic field results in a growth of clusters, and forms the so-called macro clusters. Under the influence of strong magnetic fields, well elongated macro clusters repel each other, thus forming an ordered arrangement. This cluster formation leads to the remarkable magneto-optical properties such as dichroism and birefringence. [4-8].

Optical Properties of Ferrofluids

Magnetite based ferrofluids are black and practically opaque; however thin layers exhibit the magnetic field effect both on isotropic optical properties such as extinction and anisotropic optical properties such as birefringence, dichroism etc, when they are exposed to its effect. For high concentration fluids, a difference in refractive indices of the ordinary and extraordinary rays attains 5×10^{-3} and dichroism amounts to 4×10^{-3}. That is, the magnetic fluid displays the pronounced properties of an optically single axis crystal in an external field.

Synthesis Techniques

a) Ball Milling

Ferrofluids were synthesized first by top down methods. Bulk grained ferrites/ferromagnetic particles are converted into nanosized particles in this process by continuous grinding of them in liquid medium by ball milling technique [1,2,14]. Low energy ball milling used to take longer grinding times like 120 hrs, 100 hrs, 72 hrs etc. Later High energy ball milling (with speeds higher than 500 rpm) substituted the low energy machines which significantly reduced the milling times to 3-5 hours [15]. However, the particle deformity was higher in ferrofluids synthesized by this technique and the dispersion and stability was not ideal for application purposes.

b) Pentacarbonyl Dissociation Technique

Later another technique was developed to obtain the ferrofluids by the dissociation of iron pentacarbonyl to obtain iron nanoparticles and the entire process was carried out inside vacuum chamber and the surfactant solution was sprayed on to the nanoparticles so as to obtain a wet slurry of iron nanorprticles coated by the surfactant. The slurry can be dispersed in suitable carrier liquids to obtain the ferrofluids. Other metal carbonyls like that of cobalt and nickel can also be prepared by this method. However, the limitation of this technique is that the technique can be used for the synthesis of metal based ferrofluids only[16-19].

c) Co- Precipitation+Insitu Surfactant Coating

This is the technique that is used widely for the synthesis of ferrofluids nowadays [2,3,20]. For the synthesis of organic solvent based ferrofluids, the reactants are taken in appropriate molar ratio and each one is dissolved in water separately. They are mixed together with continuous stirring and titrated against NH_4OH solution slowly and the surfactant solution is sprayed into the beaker once the particle formation is complete. The wet slurry is collected by magnetic separation and washed several times with acetone and water and finally dispersed in suitable organic carriers. Substituted ferrites like zinc, nickel, manganese, cobalt substituted ferrites can also be synthesized as ferrofluids by slight modification of the above mentioned protocol [21-26].

Ionic ferrofluids are mostly aqueous ferrofluids. For aqueous based ferrofluids, , tetra methyl ammonium hydroxide [27-29], citric acid [30] etc. are used for providing ionic repulsion. The nanoparticles are synthesized by a similar procedure as described earlier.

Applications of Ferrofluids

Ferrofluids have numerous applications. Some of them can be listed below.

a) As a Lubricants and Coolant in Motors

As ferrofluids can be driven through an applied magnetic field, a small drop of ferrofluid is just enough for serving as a lubricant in motors. This can simultaneously behave as a lubricant and coolant. One way of extracting heat from an equipment, which heats up by functioning, and so keeping it not too hot, is by using a good heat conductor which connects the equipment to some mass which has much bigger heat capacity and, perhaps, much bigger open surface to dissipate heat. In some cases the good heat conductor must not be a solid, because it would block the equipment's operation (for example, if it has to vibrate). One way to achieve the desired goal is by using a ferrofluid as heat conductor. (A non magnetic liquid would flow away from the place where it is supposed to operate). Numerous research works have been completed in this area [31-43].

b) For High Quality Leak Proof Sealing

Ferrofluids consists of tiny magnetic particles which are about 10nm or less in size. The number density is so high that in an applied magnetic field, it behaves as almost closely packed solid. Hence the most important property of ferrofluids is that they can be employed as a sealant where a 100% leak proof sealing is necessary. In hard disks of computers, which have to operate in an hermetically closed box because any grain of powder or even smoke may spoil the reading and writing process. Therefore it is necessary to seal hermetically the hole through which the axle passes. This is achieved by making the hole inside a magnet (see Figure 9) and the shaft made of soft magnetic material. A groove in the shaft is filled up with ferrofluid, which is kept in place by the magnetic field, obstructing the passage of any impurity, but leaving the axle free to rotate, because the obstructing material is liquid [1, 2].

c) In Loud Speaker Coils

Ferrofluids can serve as a lubricant and coolant in loud speaker coil. Additionally, it can enhance the velocity of sound as it serves as a magnetic fluid medium. In loudspeaker, whose coil heats up by functioning and the ferrofluid is kept in place by the magnetic field of the magnet which is fixed on the loudspeaker's horn. Nowadays most of the high power loudspeakers are equipped with ferrofluid. The presence of the fluid around the coil improves also the quality of the speaker because it damps unwanted resonances [1, 2].

Other applications of ferrofluids are listed below.

1. As a pressure, volume and temperature sensor[8, 44-47]
2. As a magnetic field and moment sensor[22]
3. As Nonlinear optical limiter [48]
4. Biomedical applications like targeted drug delivery, hyperthermia as a contrasting agent in MRI etc. [49].
5. In Magic slates
6. In magneto optical display devices

The industrial applications of ferrofluids do not end here. It has vast application potential. In future days, numerous emerging applications of these smart fluids will be made known before the public which make them to be called as "Smart Fluids".

CONCLUSION

This chapter gives a general introduction of ferrofluids, the basic physical laws governing the stability of ferrofluids, and some important properties and applications of ferrofluids. Detailed analysis of each of these properties and applications will be explained in the following chapters.

REFERENCES

[1]   S.S. Papell, U.S. Patent 3, 215, 572, 1965.
[2]   R. E. Rosenswieg; *Magnetic Fluids*, Oxford University press, p.124.
[3]   B. M. Berkovsky, V. F. Medvedev, M. S, Krakov; *Magnetic Fluids, Engineering Applications*; Oxford University Press. 1993
[4]   H.E. Horng, C.Y.Hong, S.Y. Yang, H.C. Yang, *J. Phys. and Chem. of Solids*, 62, (2001)1749.

[5]   Tengda Do, Weili Luo, *J. Appl. Phys*, 85, (1999), 5953.

[6]   S. Swapna Nair, S. Rajesh, V.S Abraham, and M.R Anantharaman, J.Magn.Magn.Mater, 304 (2006) 28.

[7]   S. Swapna Nair, Francis Xavier, P.A.Joy, S.D. Kulkarni, M.R. Anantharaman, *J. Magn. Magn. Mater.*, 320, (2008), 815.

[8]   R. V.Mehta and R.V. Upadhyay, *Curr. Scie.*,76, (1999), 305.

[9]   S. Chikazumi, S. Taketomi, M. Ukita, M. Mizukami, H. Miyajima, M. Setogawa, and Y. Kurihara; *J. Magn.Magn.Mater.* 65 (1987) 245.

[10]   K.T. Wu, Y.D. Yao, *J. of Appl. Phys*, 85, (1999) 5959.

[11]   B.D.Cullity, 2nd edn. '*Introduction to Magnetic Materials*' Addison-Wesley Publishing Company, Inc. Boston, 2007.

[12]   J.C Bakri, R. Perzynski, D. Salin, V, Cabuil and R. Massart, *J.Magn.Magn.Mater* 85(1990) 27.

[13]   Hydrodynamica, Britanica online Encyclopedia, Retrieved 2008, p. 10-30.

[14]   R. E. Rosensweig, J. W. Nestor and R. S. Timmins, *Proc. Symp. AlChE-IChemE,* Ser. 5, pp. (1965) 104 (discussion, pp. 133-137).

[15]   R.Y. Hong, Z.Q. Ren, Y.P. Han, H.Z. Li, Y. Zheng, J. Ding, *Chem. Engg. Scie.* 62, 2007, 5912.

[16]   T. W. Smith and D. Wychick, *J. Phys. Chem.* 84, (1980), 1621.

[17]   J. R. Thomas *J. Appl. Phys.* 37 (1966) 2914.

[18]   P. H. Hess, P. H. Parker, *J. Appl. Polymer Sci.,* 10, (1966) 1915.

[19]   T. W. Smith, U S Patent (1981) 4,252,673

[20]   S. E Khalafalla, G. W. Reimers, U S Patent (1973) 3,764,540

[21]   L.F. López, G. Bahamón, J. Prado, J.C. Caicedo, G. Zambrano, M.E. Gómez, J. Esteve, P. Prieto, *J. Magn.Magn.Mater*, 324, (2012), 394.

[22]   S. Swapna Nair, S. Rajesh, V. S. Abraham and M. R. Anantharaman, *Bull.Mater.Scie*, 34 (2011), 245.

[23]   V.S. Abraham, S. Swapna Nair, S. Rajesh, U.S. Sajeev and M.R. Anantharaman, *Bull. Mater. Scie*, 27, (2004), 155.

[24]   T. J. Shinde, A. B. Gadkari and P. N. Vasambekar, *J. Mater. Scie*, Article available online, DOI: 10.1007/s10854-011-0474-y.

[25]   T. J. Shinde, A. B. Gadkari and P. N. Vasambekar, *J. Mater. Scie,* 21, 2010, 120.

[26]   B. Gadkari, T. J. Shinde and P. N. Vasambekar, *J. Mater. Scie,* 21, 2010. 96.

[27]   S. Palacin, P. C. Hidber, J. Bourgoin, C. Miramond, C. Fermon, G. Whitesides, *Chem. Mater,* 8 (1996), 1316.

[28]   J. P. Jolivet, R. Massart, J. Fruchart, M. Nouv. *J. Chim.*, 7, (1983) 325.

[29]   P. Enzel, N. B. Adelman, K. J. Beckman, D. J. Campbell, A. B. Ellis, G. C. Lisensky, *J. Chem. Educ.* 76 (1999), 943.

[30]   A.P. Reena Mary , T.N. Narayanan , V. Sunny , D. Sakthikumar , Y. Yoshida, P.A. Joy, M.R. Anantharaman , *Nanoscale Res Lett.* (2010) 5 1706.

[31]   J. L. Neuringer, R. E. Rosensweig, *Phys. Fluids,* 7 (1964)1927.

[32]   S. ODENBACH (Editor), *Ferrofluids: Magnetically controllable fluids and their applications,* Lecture, Notes in Physics, Springer-Verlag, 253 pags (2002).

[33] S. W. CHARLES, The preparation of magnetic fluids, in: S. ODENBACH (Editor), *Ferrofluids: Magnetically controllable fluids and their applications,* Lecture Notes in Physics, Springer-Verlag, pp.3-18,2002.

[34] S. W. Charles, *Rom. Repts. Phys.*, 47 (1995) 249.

[35] L. Vekas, D. Bica, M. V. Avdeev, *China Particuology*, 5, 43 (2007).

[36] L. Anton, I.De Sabata, L. Vekas, *J. Magn. Magn. Mater.*, 85, (1990) 219.

[37] K. RAJ, Magnetic fluids and devices: a commercial survey, in: B. Berkovsky, V. Bashtovoi (Eds.), *Magnetic fluids and applications handbook,* Begell House, New York, pp.657-751 (1996).

[38] V. Cabuil, J.C. Bacri, R. Perzynsky, YU. Raikher, Colloidal stability of magnetic fluids, in: Berkovsky, V. Bashtovoi (Eds.), *Magnetic fluids and applications handbook,* Begell House, New York, pp.33-56 (1996).

[39] M.I. Shliomis, Ferrohydrodynamics: Retrospective and Issues, in: S. ODENBACH (Editor), *Ferrofluids: Magnetically controllable fluids and their applications,* Lecture Notes in Physics, Springer-Verlag, pp.85-110, 2002.

[40] Tarapov, *Magnetohydrodynamics*, 8, (1972) 444.

[41] R. C. Shah and M. Bhat, *Trib. Inter.*, 37, (2004) 441.

[42] Q. Zhang, S. Chen, S. Winoto, and E. Ong, *IEEE Trans. Magn.*, 37, (2001) 2647.

[43] P. Kuzhir, *Trib. Inter.*, 41, (2008) 256.

[44] C. Cotae, O. Baltag, R. Olaru, D. Calarasu, D. Costandache. *Sens. and Act.*, 84, (2000) 246.

[45] J. C. Bacri, J. Lenglet, R. Perzynski, J. Servais, *J. Magn. Magn. Mater.* 122, (1993), 399.

[46] Stanci, V. lusan, C. D. Buioca. *J. Magn. Magn. Mater.*, 201, (1999) 394.

[47] http://www.nplindia.org/ferrofluid-based-temperature-sensor

[48] S. Swapna. Nair, J. Thomas, S. Sandeep, M. R. Anantharaman and R. Philip, *Appl. Phys. Lett.* 92, (2008), 171908.

[49] J. Roger, J. N. Pons, R. Massart, A. Halbreich, J. C. Bacri, *Eur.Phys, J.Appl.Phys*, 5, (1999) 321.

See also

http://www.directindustry.com/prod/ferrolabs/ferrofluid-based-seals-30165-136078.html

http://machinedesign.com/article/magnetic-fluids-tackle-tough-sealing-jobs-0217

http://liquidsresearch.com/en-GB/for_sealing_applications-52.aspx

http://www.sdmaterials.com/ferrofluid_seals.html

http://www.ferrotec.com/products/ferrofluidic/

http://www.directindustry.com/prod/ferrolabs/ferrofluid-based-seals-30165-136078.html

In: Ferrofluids
Editors: Franco F. Orsucci and Nicoletta Sala

ISBN: 978-1-62808-410-8
© 2013 Nova Science Publishers, Inc.

Chapter 2

MAGNETISM: GENERAL IDEAS

Swapna Nair[*][1] *and M. R. Anantharaman*[2]

[1]Departmento de Engenharia Cerâmica e do Vidro and CICECO,
Universidade de Aveiro, Aveiro, Portugal
[2]Department of Physics,
Cochin University of Science and Technology Kochi, India

ABSTRACT

This chapter briefly discusses the theoretical concepts of magnetism. The initial discussion is about magnetism, origin of magnetism in materials, introduction to various magnetic terms etc. while the latter sections details the magnetic properties exhibited by nanomagnetic properties which are necessary to learn the magnetic properties of ferrofluids.

INTRODUCTION

To introduce the general ideas of magnetism, it is necessary to learn primarily about various types of magnetism like diamagnetism, paramagnetism, ferromagnetism, antiferromagnetism and ferrimagnetism.

DIAMAGNETISM

When a molecule is subjected to a magnetic field those electrons in orbit planes at a right angle to the field will change their momentum as predicted by Faraday's Law which tells us that when the applied magnetic field is increased there will be a an induced E-field which acts in such a way as to oppose the applied field (Faraday and Lenz Law).

---

[*] E-mail: swapna.s.nair@gmail.com.

This behaviour, whereby the induced moment is opposite to the applied field, is present in all materials and is called diamagnetism. Hydrogen, copper, graphite etc. are examples for diamagnetic substances. Diamagnetic materials are those whose atoms have only paired electrons.

Paramagnetism

Paramagnetic materials have unpaired electrons. An O_2 molecule has a net or permanent magnetic moment even in the absence of an externally applied field. If an external magnetic field is applied, the 'poles' of the molecule tend to line up parallel with the field and reinforce it. Such molecules, with permanent magnetic moments are called paramagnetic.

Although paramagnetic substances like oxygen, tin, aluminium and copper sulphate are attracted to a magnet the effect is almost as feeble as diamagnetism. This is due to the thermal randomisation of the spin alignment.

Ferromagnetic Materials

The most important class of magnetic materials is the ferromagnets: Typical examples are iron, nickel, cobalt and manganese, or their compounds (and a few rare earths and rare earth based compounds). The magnetization curve looks very different to that of a diamagnetic or paramagnetic material.

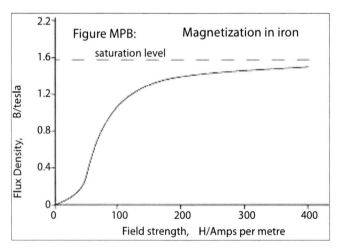

Figure 1. Magnetisation curve of a ferromagnetic material.

At an atomic level ferromagnetism is explained by a tendency for neighbouring atomic magnetic moments to become locked in parallel with their neighbours. This is only possible at temperatures below the *curie point*, above which thermal disordering causes a sharp drop in permeability and degeneration into paramagnetism. Ferromagnetism is distinguished from paramagnetism by more than just permeability because it also has the important properties of remanance and coercivity.

Antiferromagnetic Materials

In materials that exhibit antiferromagnetism, the magnetic moments of atoms or molecules, align in a regular pattern with neighboring moments (on different sublattices) pointing in opposite directions. This is again a manifestation of ordered magnetism. However, as the alignment of spins is antiparallel, net magnetic moment is much lower than that is exhibited by the ferromagnetic materials. Generally, antiferromagnetic order may exist at sufficiently low temperatures, vanishing at and above a certain temperature called the Néel temperature (named after Louis Néel, who had first identified this type of magnetic ordering). [L. Néel, Propriétées magnétiques des ferrites; Férrimagnétisme et antiferromagnétisme, Annales de Physique (Paris) 3, 137–198 (1948)]. Above the Néel temperature, the material is typically paramagnetic.

Ferrimagnetic Materials

Ferrites are ferrimagnetic materials are ideal candidates for most of the applications as they possess permeability to rival most ferromagnets but their eddy current losses are far lower than ferromagnets because of their high electrical resistivity. Almost every item of electronic equipment produced today contains some ferrites. Also it is easy to mould by pressing or extruding - both low cost techniques Loudspeakers, motors, deflection yokes, interference suppressors, antenna rods, proximity sensors, recording heads, transformers and inductors are frequently made based of ferrites.

At the atomic level, the magnetic properties depend upon interaction between the electrons associated with the metal ions. Neighbouring atomic magnetic moments become locked in anti-parallel with their neighbours (just like antiferromagnets). However, the magnetic moments in one direction are weaker than the moments in the opposite direction leading to an overall magnetic moment.

Magnetic Hysteresis (M-H Curve)

When a ferromagnetic material is magnetized in one direction, it will not relax back to zero magnetization when the imposed magnetizing field is removed. It must be driven back to zero by a field in the opposite direction.

If an alternating magnetic field is applied to the material, its magnetization will trace out a loop called a hysteresis loop. The lack of retraceability of the magnetization curve is the property called hysteresis and it is related to the existence of magnetic domains in the material. Once the magnetic domains are reoriented, it takes some energy to turn them back again. This property of ferrromagnetic materials is useful as a magnetic "memory". Some compositions of ferromagnetic materials will retain an imposed magnetization indefinitely and are useful as "permanent magnets". The magnetic memory aspects of iron and chromium oxides make them useful in audio tape recording and for the magnetic storage of data on computer disks.

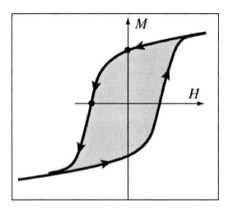

Figure 2. Magnetic hysteresis loop of a ferromagnetic material.

Hysteresis loop (M-H curve) of a magnetic material is provided in Figure 2. The magnetisation is plotted against the applied magnetic field. It can be seen that there is a loop with a definite area which is proportional to the energy loss in one magnetisation cycle.

SATURATION MAGNETISATION (M_S) COERCIVITY (H_C) AND REMENANCE (M_R)

A good permanent magnet should produce a high magnetic field with a low mass, and should be stable against the influences which would demagnetize it. The desirable properties of such magnets are typically stated in terms of the remanence and coercivity of the magnet materials. Saturation magnetisation, Remanence and coercivity of a magnetic material can be determined from the hysteresis curve (M-H curve).

Saturation magnetisation is the maximum magnetisation attained by the material upon the application of an external magnetic field. After attaining saturation, the magnetisation will remain almost steady upon further application of an external magnetic field. The saturation magnetisation of a material can be determined from the hysteresis curve.

Remanence and coercivity are other important properties of magnetic materials which is to be determined before their employment in various applications.

As already explained, when a ferromagnetic material is magnetized in one direction, it will not relax back to zero magnetization when the imposed magnetizing field is removed. The amount of magnetization it retains at zero driving field is called its remanence (See Figure 1). It must be driven back to zero by a field in the opposite direction; the amount of reverse driving field required to demagnetize it is called its coercivity.

Magnetic Susceptibility and Permeability

The magnetic susceptibility χ_m is a dimensionless proportionality constant that indicates the degree of magnetization of a material in response to an applied magnetic field. It can be shown using the relation,

$$M= \chi_m H \tag{1}$$

M is the degree of magnetisation and H is the applied magnetic field. It can be obtained as the slope of the linear portion of the M-H curve.

Magnetic flux density B can be related to M and H by a relation

$$B= \mu 0 \,(H+M) \tag{2}$$

$$= \mu 0 \,(H+ \chi_m H) \tag{3}$$

$$= \mu 0 \,H(1+ \chi_m) \tag{4}$$

Magnetic flux density B and the applied magnetic field H can be related by a relation

$$B= \mu H \tag{5}$$

where μ is the effective permeability of the material. The effective permeability can be related to the absolute permeability μ_0 as

$$\mu = \mu_0\mu_r \tag{6}$$

where μ_r is the relative permeability of the material. From (4), (5), and (6), it can be concluded that

$$\mu_r=(1+ \chi_m) \tag{7}$$

Hence relative permeability and susceptibility are closely related proportionality constants.

Magneto Crystalline Anisotropy

Magnetization curves for single crystal iron at $18°C$ are shown in figure. It can be seen that, along the [100] direction, a very small magnitude of the magnetic field is required to produce a large magnetization; in other words it is 'easy' to magnetize an iron crystal along [100] direction, while it is 'hard' to do so along the other directions.

The preference of spontaneous magnetization for a specific, easy direction may be taken as an evidence of a thermodynamically stable, minimum energy state. Now, if an external magnetic field is applied so as to rotate the magnetization vector out of its easy direction the energy of the system will tend to increase. This increase in energy may be called the magneto crystalline energy. If 'θ' is the angle between the easy direction and the magnetization vector the anisotropy energy will be zero for $\theta=0$, and it will increase with increase in 'θ' values. While the exchange energy is isotropic in nature, since it is determined by the relative directions of the adjacent magnetic spins only, the anisotropy energy is believed to be due to spin orbit coupling. As a result of this L-S coupling in the presence of a crystal field, atomic spins align themselves along a preferred direction with respect to the unit cell (crystal) axes. When an external magnetic field rotates the spins, the residual, unquenched orbital magnetization vector also rotates with the spin magnetization. This rotation alters the orbital overlap of the adjacent magnetic atoms or ions.

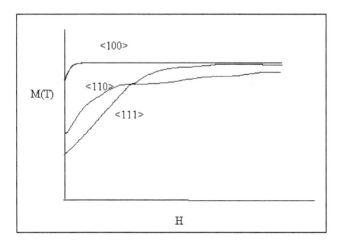

This change in the overlap, in turn, affects electrostatic coulombic and the exchange energies. As a consequence, the total energy of the system increases. For cubic crystals, the anisotropy energy may be expressed as:

$$E = K_1(\alpha_1^2 \alpha_2^2 + \alpha_2^2 \alpha_3^2 + \alpha_3^2 \alpha_1^2) + K_2 \alpha_1^2 \alpha_2^2 \alpha_3^2 \qquad (8)$$

where K_1, K_2, are constants and $\alpha_1, \alpha_2, \alpha_3$ are the direction cosines of the magnetization vector with respect to the unit cell axes. For uniaxial crystals such as Cobalt,

$$E = K_1 \sin^2 \theta + K_2 \sin^4 \theta, \qquad (9)$$

where θ is the angle between the easy and the magnetization directions [1-6].

Exchange Interactions

J.Frenkel made the assumption that the origin of strong, ferromagnetic interactions is due to what are called as exchange interactions. It is a characteristic quantum effect having no parallel in classical physics. Given below is a brief account of the various types of exchange interactions present in solids [2-4].

Direct-Exchange Interaction

This mechanism is operative between the adjacent atoms/ions and leads to a strong magnetic coupling. The total energy of the system corresponds to two situations:

a) Symmetric state corresponding to antiparallel spins
b) Antisymmetric state corresponding to parallel spins.

The corresponding energies can be described by:

$$E_A = 2E_O + \frac{E_C + E_{ex}}{1 + S^2} \tag{10}$$

$$E_B = 2E_O + \frac{E_C - E_{ex}}{1 - S^2} \tag{11}$$

Here $2E_O$ is the total energy of the isolated atoms (A, B), E_C the Coulombic interaction between electrons, nuclei and electrons-nuclei, E_{ex} the exchange energy associated with the process of exchanging electrons and S the overlap integral, whose value lies between 0 and 1. The magnitude of E_{ex} is always much larger than that of E_C and, therefore, the stability of the symmetric/antisymmetric state depends on the term E_x. In turn, E_{ex} depends on two factors:

1. The dot product, $Si.Sj$ where Si and Sj are the total spins of the adjacent atoms.
2. The exchange integral J_{ex}, which represents the probabilities of exchange of electrons.

J_{ex} is obviously a sensitive function of overlap of electronic wave functions and its sign can be either positive or negative.

If $J_{ex} > 0$, parallel alignment of the neighboring spins is favored and this corresponds to the ferromagnetic case. If $J_{ex} < 0$, neighboring spins will align themselves antiparallel, corresponding to antiferromagnetic behavior [2-4].

Indirect or Super-Exchange Interaction

The value of the exchange integral depends on the ratio of the inter-atomic distance (a) to the diameter of the 3d orbital (D). The ferromagnetism of Fe, Co and Ni can be attributed to this direct exchange.

As the degree of overlap of d–orbitals decreases, J_{ex} becomes positive. This is true for values of $a/D > 1.5$. If $a/D < 1.5$, indirect exchange interaction leads to antiparallel alignment of adjacent atomic magnetic moments, resulting in the antiferromagnetic ordering [1-4].

Magnetism at the Nano Level

At the nanolevel the properties of the materiall changes from that of the bulk. Some of the interesting phenomena are explained below [2, 4, 7].

Coercivity of Fine Particles

In magnetic studies on fine particles the single property of most interest is the coercivity, for two reasons:

1. It must be high, at least exceeding a few hundred Oersteds, and
2. It is a quantity which comes quite naturally out of theoretical calculations of the hysteresis loop.

The coercivity of fine particles has a striking dependence on their size. As the particle size is reduced, it is typically found that the coercivity increases, goes through a maximum, and then tends towards zero.

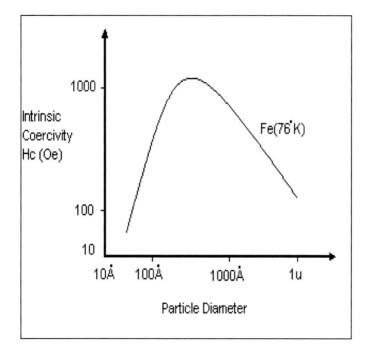

Figure 3. Variation of coercivity with particle size.

Beginning at large sizes, we can distinguish the following regions:

1. Multidomain: magnetization changes by domain wall motion. For most materials the size dependence of the coercivity is experimentally found to be given approximately by

$$H_{ci} = a + \frac{b}{D} \qquad (12)$$

where a and b are constants

2. Single-domain: Below a critical distance D_s, which is not well defined, the particles become single domains, and in this size range the coercivity reaches a maximum.

Particle of size D_s and smaller change their magnetization by spin rotation, but more than one mechanism of rotation can be involved.

a) As particle size decreases below D_s the coercivity decreases, because of thermal effects, according to

$$H_{ci} = g - \frac{h}{D^{3/2}} \tag{13}$$

where g and h are constants.

b) Below a critical diameter D_p the corecivity is zero, again because of thermal effects, which are now strong enough to spontaneously demagnetize a previously saturated assembly of particles. Such particles are called superparamagnetic.

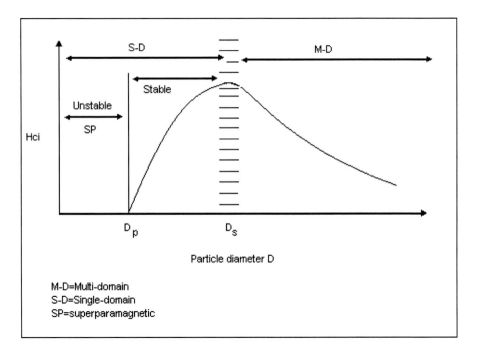

Figure 4. Variation of intrinsic corecivity H_{ci} with particle diameter D (schematic).

Superparamagnetism

In ferromagnetic materials there is spontaneous magnetization, which arises due to the interaction between the neighbouring atomic magnetic dipoles. It is called spin exchange interaction and is present in the absence of external magnetic field. The exchange interaction aligns the neighbouring magnetic dipole moments parallel to one another and this spreads over a finite volume of the bulk. This small volume is called the domain. Each domain is spontaneously magnetized, the magnetization being appropriate to temperature T. In an unmagnetised piece of ferromagnet the domains are not aligned. When external field is applied magnetization of the specimen may occur either by the growth of one domain at the expense of another i.e. by the motion of domain walls.

If the size of the ferromagnetic particle is reduced below a critical particle size it would consists of single magnetic domain. This single domain particle is in a state if uniform magnetization at any field. Let us consider such a particle whose total magnetic moment is directed at an angle θ to an applied field H.

For the sake of simplicity let us consider only one preferential direction (direction of easy magnetization) and let us call V the particle volume and θ the angle between the easy axis and the magnetic moment directions.

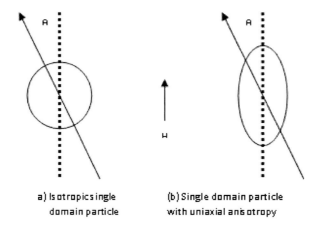

a) Isotropic single domain particle

(b) Single domain particle with uniaxial anisotropy

Figure 5. Single domain Particles in magnetic field.

The anisotropy energy,

$$E = KV\sin^2\theta \qquad (14)$$

K is also sometimes called anisotropy constant but one must keep in mind K may depend drastically on temperature.

Consider an assembly of aligned uniaxial particles that are fully magnetized along the easy symmetry axis. After the field is removed the resulting remanance will vanish as,

$$M_r = M_s \exp\left(-\frac{t}{\tau}\right) \qquad (15)$$

M_s is the full magnetization
t is the time after the removal of field
τ is the relaxation time for the process
The relaxation time is given by

$$\frac{1}{\tau} = f_0 \exp\left[-\frac{KV}{kT}\right] \qquad (16)$$

As the particle volume V becomes smaller, the relaxation rate increases. Hence let us consider an observation time τ_m, characteristic of the measurement technique. If $\tau_m \gg \tau$, the measurement result is averaged over a great number of reversals. For example under zero fields the magnetic moment of a particle is averaged to zero. This is the superparamagnetic state.

If $\tau m \ll \tau$, the magnetic moment appears blocked in one of the two directions of the easy axis. This is the blocked state. Hence, depending on the values of the anisotropy constant, the particle volume and the characteristic measurement time, it may be possible to evidence the transition from the superparamagnetic to the blocked regime by decreasing the temperature. The temperature at which this transition occurs is called the blocking temperature [2, 4, 8].

However the two main aspects of superparamagnetism are

1. Magnetisation curves measured at different temperatures superimpose when M is plotted as a function of H/T.
2. There is no hysteresis, i.e., both the retentivity and corecivity are zero.

Thermal Equilibrium Properties of Single Domain Particles

Consider a particle whose total magnetic moment μ is directed along an angle θ to an applied field H. If there are no anisotropic terms in the energy, the energy of this particle is -$\mu H \cos\theta$. In nature, single domain particles are not fully isotropic in their properties, but will have anisotropic contributions to their total energy associated with the external shape of the particle, imposed stresses or the crystal structure itself. Consider an anisotropy that is uniaxial in symmetry of the form: $E_k = KV \sin^2\theta$. Here θ is the angle between the moment and the symmetry axis of the particle, V is the volume of the particle and K is the anisotropy energy per unit volume. For an applied field H along the symmetry axis, the energy of the particle is given by: $KV \sin^2\theta - \mu H \cos\theta$. Here we have a different Boltzmann distribution of θ's in thermal equilibrium than we had with out the anisotropy term and the magnetization curve will no longer be a simple Langevin function [2,4].

Approach to Thermal Equilibrium

Consider an assembly of aligned uniaxial particles that are first fully magnetized along the easy symmetrical axis On the removal of the field, the resulting remanance will vanish as $M_r = M_s \exp(-t/\tau)$ where M_s is the saturation magnetization, t is the time after removal of the field and τ is the relaxation time for the process, given by

$$\frac{1}{\tau} = f_0 \, exp \, (-KV/kT), \tag{17}$$

where f_0 is the frequency factor of the order of 10^9 sec^{-1}. To approach the thermal equilibrium, a sufficient number of particles should be reversed by thermal activation over the energy barrier KV, the probability of which process is proportional to exp (-KV/kT).

For particles with an anisotropy of cubic symmetry, the energy barrier between adjacent easy directions will appear in the exponential. Using standard definitions of the first order cubic anisotropy constant, the barrier is KV/4 for K>0.([100] easy direction).

Because of the exponential dependence of τ on particle volume, there is a fairly well defined particle size at which the transition to stable behaviour occurs. There will be in general, only a narrow range of particle sizes in which the relaxation times, for which measurable "magnetic viscosity" effects would be expected. As a rough measure of the size corresponding to $\tau = 10^2$ sec. This occurs when the energy barrier is equal to approximately 25KT. For a given particle, the temperature at which this occurs has been called the "blocking temperature".

Consider an assembly of uniaxial particles which is in an initial state of magnetization M_i by an applied field. Now the field is reduced to zero at a time t=0. Some particles in the assembly will reverse their magnetization as their thermal energy is larger and the magnetization of the assembly tends to decrease. The rate of decrease at any time is proportional to the magnetization existing at that time and the Boltzmann factor.

Therefore

$$-\frac{dM}{dt} = f_0 M e^{\frac{-KV}{kT}} = \frac{M}{\tau} \tag{18}$$

The proportionality factor is called the frequency factor and it has a value of 10^9sec^{-1}.
The constant τ is called the relaxation time.
Rearranging the above equation and integrating we arrive at

$$M_r = M_i e^{\frac{-t}{\tau}} \tag{19}$$

and hence relaxation time τ can be defined as the time for remanence M_r to decrease to 1/e of its initial value. From equations (17 and 18), we can write

$$\frac{1}{\tau} = f_0 e^{\frac{-KV}{kT}} \tag{20}$$

Thus it is clear that τ is strongly dependent on V and T.
If we put relaxation time as 100sec, then from equation (18,19, and 20) we arrive at

$$\frac{KV}{kT} = 25 \tag{21}$$

Hence transition to stable state occurs when energy barrier equals 25kT and also the upper limit of the particle volume for Superparamagnetism for uniaxial particles is given by

$$V_p = \frac{25kT}{K} \tag{22}$$

For a particle assembly of constant size there will be particular temperature called the superparmagnetic blocking temperature, below which magnetization will be stable. For a uniaxial particle assembly,

$$T_B = \frac{KV}{25k} \tag{23}$$

Now we can consider the effect of an applied field on the approach to equilibrium. Assume an assembly of particles with their easy axis parallel to the z axis. Let it be initially saturated in the +z direction and let us now apply a field in the −z direction. Then the Ms in each particle will make an angle θ with the z axis. Then the total energy per particle is

$$E = V(K\sin^2\theta + HM_s\cos\theta) \tag{24}$$

The energy barrier for reversal is the difference between the maximum and minimum values of E and it can be obtained as

$$\Delta E = KV(1 - \frac{HM_s}{2K})^2 \tag{25}$$

The particles with sizes larger than D_p are stable in zero filed and will not thermally reverse in 100s. But when the field is applied the energy barrier can reduce to 25kT. This will be the coercivity and is given by

$$\Delta E = KV(1 - \frac{H_{ci}M_s}{2K})^2 = 25kT \tag{26}$$

$$H_{ci} = (\frac{2K}{M_s})\left[1 - (\frac{25kT}{KV})^{1/2}\right] \tag{27}$$

when V is very large or T is zero $H_{ci} = (2K/M_s)$

Putting this limiting value as

$H_{ci,0}$ and substituting for $25k_B T/K$ as V_p, we get

$$h_{ci} = \frac{H_{ci}}{H_{ci,0}} = 1 - (\frac{V_p}{V})^{1/2} = 1 - (\frac{D_p}{D})^{3/2} \tag{28}$$

Similarly equation (28) can also be used for variation of coercivity with temperature for particles of constant size. For this we assume that particle with critical size V_p have zero coercivity at their blocking temperature T_B.

We get

$$h_{ci} = \frac{H_{ci}}{H_{ci,0}} = 1 - (\frac{T}{T_B})^{1/2} \tag{29}$$

Particles larger than V_P have finite retentivity as thermal energy cannot reverse their magnetization in 100s. To find a relation between retentivity and size we can combine equations (29) and (30) to get

$$\ln\frac{M_r}{M_i} = -\frac{t}{\tau} = -10^9 e^{-KV/kT}$$

$$(30)$$

which gives the grain size dependence of magnetic properties which varies with temperature [2,4].

Spin-Glass and Spin Glass Clusters

Crystalline alloys comprising of a small percentage of the magnetic ions in a nonmagnetic matrix show some peculiar magnetic behaviour when cooled below a critical temperature T_f, the spins can order speromagnetically and this effect is named as spin glass freezing typical examples being Fe or Mn in Cu or Au matrix. A characteristic feature of the so called spin glass, is that a cusp is seen in the low field susceptibility - temperature curve at T_f. As the constituent elements in glass are frozen randomly, the spins are frozen at random in this alloy and hence the name spin-glass.

For a single domain particle, below a critical volume, the energy barriers become comparable to the thermal energy and thus a total magnetic moment of the particle can fluctuate between the easy directions. In real systems the moments of the particles interact between them and there is always a particle size and shape distribution, as well as distribution of particle environments, depending on the topology of the system. This leads to a distribution of the total energy barriers of particles, because of the different values of the various contributions and thus to a distribution of the blocking temperature T_B. At enough low temperatures ($T<T_{Bmin}$), below which all the particles moments are blocked along their anisotropy axes, (small fluctuations around them are still possible: collective magnetic excitations) a disordered magnetic arrangement will result, recalling the frozen disordered magnetic state of spin-glass systems.

Neel superparamagnetic model was proposed to explain the macroscopic properties of the spin glass state which implies the existence of a distribution of magnetic clusters, the passage from the paramagnetic to spin-glass state is described in terms of a progressive blocking of clusters moments [2].

Surface Magnetism

Fine particles provide an attractive platform for the study of the magnetic properties at the surface. They possess variety of advantages over the thin films. Their Surface to volume ratio is very high and can be varied over a relatively wide range. It may be possible to distinguish between the surface, close to surface and bulk properties. Also fine particles have only one interface while, thin films have two and at least one of these, contact with the substrate.

Interface of fine particles are of vacuum, gas, liquid or solid surfactant, or the contact with a binder. Fine particles may be superparamagnetic, a phenomenon that depends on both the volume and the anisotropy. Any change in the surface anisotropy can be monitored by superparamagnetism measurements. They are widely employed in a number of applications, for example magnetic fluids, particulate magnetic-recording media, and catalysis in which the surface properties may be important.

The determination of the saturation magnetization M_s, of fine particles is one way of seeking information on the magnetic structure near the surface. A saturation magnetization less than the bulk value suggest that a change in the dipole arrangements has occurred. Mossbauer spectroscopy provides a rather direct way for the investigation of magnetic structures.

Fine particles can also contribute to determination of the hyperfine field magnitude at the surface. The chemical composition or the morphology is the most important factor influencing the non-collinear magnetic structure in fine particles. Particle morphology is important, and must be considered in any detailed analysis of surface magnetism [8-10].

CONCLUSION

The discussion of different types of magnetism and the effect of particle size on the magnetic properties of materials are quite important in understanding the magnetic properties of ferrofluids. The latter section of the chapter dedicates to the special magnetic properties and phenomena happening at the nanolevel which is directly applicable to ferrofluids. This the chapter serves as a basic platform for understanding the properties explained in the coming chapters.

REFERENCES

[1]   C. Radhakrishnamurty, (1993) *Magnetism and Basalts*, Geological Society of India, Banglore, 208 pp.

[2]   B. D. Cullity, (1972) *Introduction to Magnetic Materials* Addison-Wesley Publishing Company, Inc 666 pp.

[3]   S. Chikazumi, (2002)*Physics of magnetism* , John Wiley and sons, Inc., New York

[4]   D. Jiles, (2001) *Introduction to magnetism and magnetic materials*, Chapman and hall Inc., London 536 pp.

[5]   D. Hadfield, (1962) *Permanent Magnets and Magnetism*, John Wiley and sons, Inc., London 430 pp.

[6]   L. Neel. *Ann. Phys.* 31(948), 137.

[7]   S. Gangopadhyay, G. C. Hadjipanyas, C. M. Soresen and K. J. Klabunde, (1994) *Nanophase Materials – Synthesis properties and applications* (G. C. Hadjipanayis and R.W. Seigel (Edt.)), Kluwer Academic Publishers 573.

[8]   Qi Chen, Adam. J. Rondinone, Bryan. C. Chakoumakos, Z. John. Zhang (1999) *J.Magn.Magn.Mater* 194,

[9] J. C. Ho, H. H. Hamdeh, Y. S. Chen, S. H. Lin, Y.D. Yao, R. J. Willey, S. A. Oliver *Phys. Rev. B* 52 (1995) 10122.

[10] H. Hamdeh, J. C. Ho, S. A.Oliver, R. J. Willey, G. Oliveri, G. Busca *J. Appl. Phys.,* 81 (1997) 1851.

In: Ferrofluids
Editors: Franco F. Orsucci and Nicoletta Sala

ISBN: 978-1-62808-410-8
© 2013 Nova Science Publishers, Inc.

Chapter 3

FERROFLUIDS - SYNTHESIS TECHNIQUES

Swapna Nair[1,] and M. R. Anantharaman[2]*
[1]Departmento de Engenharia Cerâmica e do Vidro and CICECO,
Universidade de Aveiro, Aveiro, Portugal
[2]Department of Physics,
Cochin University of Science and Technology Kochi, India

ABSTRACT

The focal themes of the preceding chapters were introduction to ferrofluids, details of their stability criteria, general properties, magnetic properties etc. In this chapter, different experimental techniques that are being used for the synthesis of ferrofluids are detailed. Synthesis techniques like simple wet grinding, high energy ball milling, chemical co-precipitation, reverse micelle technique, penta-carbonyl dissociation etc are introduced and explained in detail so as to enable the readers to make these fluids as their own.

INTRODUCTION

In most technological and biomedical applications of ferrofluids, the magnetic material may be one of a number of different ferrites. By far the most commonly used ferrites are magnetite (Fe_3O_4) and maghemite ($ã-Fe_2O_3$).

Because magnetite can be oxidized to maghemite with only a relatively small reduction in moment, the actual structure of the particles in commercial and other ferrofluids usually involves the presence of both ferrites in an undefined ratio. Detailed reviews of the structure and magnetic properties of the normal and inverse spin structures of ferrites have been conducted [1-3].

The name spinel is given to those ferrites which have the formula MFe_2O_4 (where M is a divalent ion) and which have the cubic crystal structure of the mineral spinel ($MgAl_2O_4$) [1].

---

[*] E-mail: swapna.s.nair@gmail.com.

The oxygen atoms are arranged in layers in such a way that there are two types of interstitial sites, tetrahedral (A) sites and octahedral (B) sites.

The net magnetic moment of each ferrite is determined by the moment of each cation, the arrangement of the cations in the A and B sites, and the interaction between cations. In magnetite, the moment of the two Fe_{3+} ions are split between an A site and a B site and are antiferromagnetically coupled so that the moments cancel, whereas the Fe2+ ion situated on the B site gives rise to the overall moment. This opposite and unequal arrangement of the moments gives rise to ferrimagnetism. Further details are provided in chapter 2.

Synthesis of Ferrofluids

There are two basic steps in creating a ferrofluid: synthesis of the magnetic solid, magnetite (Fe_3O_4), and suspension in water with the aid of a surfactant. The magnetic particles must be very smaller, of the order of 10 nm (100 Å) in diameter so that the thermal energy of the particles is large enough to overcome the magnetic interactions between particles.

If the particles are too large, magnetic interactions will dominate and the particles will agglomerate.

Top Down Methods

Magnetic fluids were initially synthesised by top down methods [4]. This involved synthesis of micron sized magnetic counter parts through the solid state synthesis process (mixing oxides in wet medium followed by pre-sintering at a temp~500 C and finally sintering at a high temperature over 1000 C) and then grinding them in a wet medium in roll mill for one to two weeks to obtain the nanosized counterparts which could be suspended in a liquid medium.

Often the grinding process was done in the wet medium with the desired surfactant and finally the wet slurry is used to suspend in the carrier liquid of our choice. Centrifugation was usually employed to remove larger particles which could lead to agglomeration and sedimentation.

The larger time and efforts for this synthesis technique made scientists seek after other alternatives for the synthesis of ferrofluids.

High Energy Ball Milling (HEBM)

High energy ball milling is an advanced method of doing the wet grinding. Particle size reduction can be carried out by employing High Energy Ball Milling machines [5]. In this, 800 to 1500 rotations per minute can be achieved and hence the momentum imparted to the particles will be very high.

The ball to powder ratio can be maintained at 1:8 or 1:10 to impart a high momentum to these fine particles. This will enable the required size reduction within two to three hours. This results in excellent grinding performance at considerably shorter grinding times.

However there can be problems like surface defects, cation redistribution, broader grain size distribution (due to the large local heat generation) and sometimes chances of contamination of the ball/vial materials.

Another disadvantage was that only oxide based ferrofluids could be synthesised by this technique. As metals are more prone to oxidation in continuous milling, the method cannot be used.

BOTTOM UP METHODS

Chemical Precipitation Methods

The magnetite can be synthesized by a precipitation reaction that occurs upon mixing $FeCl_2$ and $FeCl_3$ with ammonium hydroxide (an aqueous solution of ammonia, NH_3) [6-8]

The unbalanced equation for this reaction is as follows:

$$\_\_ FeCl_3 + \_\_ FeCl_2 + \_\_ NH_3 + \_\_ H_2O \oslash \_\_ Fe_3O_4 + \_\_ NH_4Cl$$

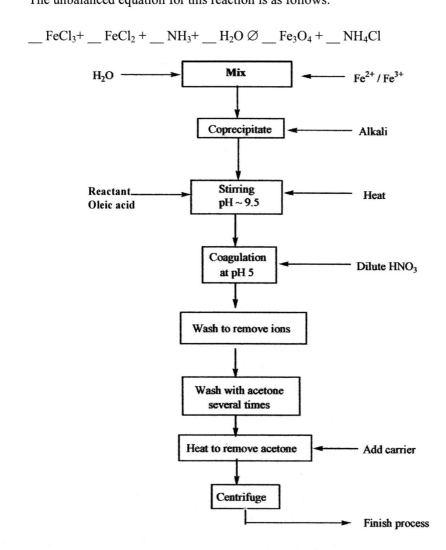

The initial reactants are anhydrous ferric chloride and ferrous chloride taken in a molar ratio 2:1. They should be dissolved in water separately and mix together thoroughly using a mechanical stirrer. Please take care not to use a magnetic stirrer for this purpose. The reason for this is explained in the proceeding steps. The precipitation of all the magnetic materials will occur in high pH of the order of pH=8 and above. Now our mixed solution containing Fe^{3+} and Fe^{2+} ions will be at a pH = 3 or below (acidic pH). Hence, this solution should be titrated against an alkali to make the pH higher to initiate the precipitation. (Both NaOH and NH_4OH can be employed. However, NH_4OH hold a better advantage as the by-product NH_4Cl can be decomposed at low temperatures itself to gaseous products. If we are using NaOH, care should be taken to wash off all the NaCl forming towards the end). Careful monitoring of the pH should be done and the surfactant should be added at a pH of 10 and stir well. The addition of alkali should be stopped just before the addition of the surfactant. In order to coagulate the surfactant coated nanoparticles, dil HCl should be added slowly till the coagulation is complete. Further addition of HCl can result in the removal of the surfactant coating. Now the wet slurry with surfactated magnetic nanoparticles is ready. The slurry should be washed several times with deionised water and then with acetone and finally the slurry can be dispersed in the organic liquids like kerosene, petroleum, hexane, toluene etc. However, alcohol cannot be used as the basefluid if the surfactant used is oleic acid.

The use of a magnetic stirrer is not advisable just because the formed magnetic nanoparticles will strongly adhere to the fish magnet we use for stirring them and the surfactation will not be perfect which results in cluster formation which is not good for the stability point of view.

Substituted Ferrites

Substituted ferrites [1,9] can also be synthesised by the method described above [10-14].

Oxalate co-precipitation is also employed recently for the synthesis of mixed ferrites and ferrofluids based on them [15-17]. However, in some cases the precipitate needs to be hydrothermally aged to facilitate the conversion of the precipitated hydroxides to the ferrite [18]. There are soft and hard ferrites. If the application demands the synthesis of a fluid with high magneto crystalline anisotropy and coercivity, cobalt ferrites are a good choice [19]. For heat transfer and thermo magnetic studies, ferrites with lower Neel temperatures can be employed as they possess high thermo magnetic coefficients [20]. For transformer and microwave applications, nickel ferrites can be used [21]. Hence ferrofluids can be synthesised with any ferrites by slightly varying the synthesis conditions explained above.

Reverse Micelle Techniques

Numerous papers have been published on the preparation of nano-sized particles including ferrites by employing reverse micelle technique. A reverse micelle can be described as a water-in-oil microemulsion in which two immiscible liquids are stabilized by a surfactant [22] or it is a spheroidal aggregate formed when a surfactant is dissolved in an organic solvent. This method involves the preparation of two microemulsions, one containing an aqueous solution of a metal salt or mixture of metal salts and the other an aqueous solution of an alkali [23,24], and mixing the two in the appropriate ratio. Using pure surfactants, the micelle-size distribution is very narrow, with the result that the particles themselves also possess a narrow particle size. For more synthesis details please refer [24, 25] The formed ferrites can be surfactated and dispersed in the carriers of choice.

Synthesis of Aqueous Ferrofluids

So far we have been discussing the synthesis procedure for the synthesis of ferrofluids with organic basefluids like kerosene, petroleum etc. However, they are not suitable for any biomedical application and hence aqueous ferrofluids emerges as a suitable candidate for these applications. There are several synthesis techniques for producing an aqueous ferrofluids. Variety of surfactant/dipersants are employed to synthesise such a fluid like tetra methyl ammonium hydroxide, citric acid, dextran etc.

Aqueous Ferrofluids with Tetramethylammonium Hydroxide ($N(CH_3)_4OH$)

The following is the synthesis scheme for aqueous ferrofluids stabilised by tetramethylammonium hydroxide ($N(CH_3)_4OH$).

The hydroxide (OH^-) ions formed in solution tend to bind to the iron sites on the magnetite particles, creating a net negative charge on each particle [26-28]. The positively-charged tetramethylammonium ions will then associate with the negatively-charged magnetite particles, forming a kind of shell around each magnetite particle. This charged shell raises the energy required for the particles to agglomerate, stabilizing the suspension.

Step by Step Procedure

$FeCl_2$ and $FeCl_3$ solutions are to be mixed in a molar ratio of 1:2M . This solution is to be titrated against NH_4OH solution till the precipitation of magnetite as described in the section *Chemical precipitation methods*. Let the magnetite settle for a few minutes. The decant is to be thrown out and the magnetite solids can be collected with the help of a magnet which is to be rinsed thoroughly several times and decant is to be thrown out in each step. 2.0 mL of 25% tetramethylammonium hydroxide is to be added at this point with thorough stirring with a glass rod. Ultrasonification also can be done at this stage to enhance dispersion. Remove the clear liquid by holding a magnet under the beaker. Sometimes, one (or some drops) drop of water is to be added to get a clear spiking fluid.

Aqueous Ferrofluids with Citric Acid for Bio Medical Applications

Monodispersed iron oxide particles of average size 9.5 nm were synthesized through controlled chemical co-precipitation method. For this, anhydrous ferric chloride ($FeCl_3$) and ferrous sulphate hepthahydrate ($FeSO_4.7H_2O$ from Merk) in the molar ratio of 2:1, each in 500ml of distilled water were taken as the precursor solution. 12% of NH_4OH was added to the solution while stirring at room temperature to supersaturate for the precipitation of the oxide. The rate of reaction was controlled by allowing one drop of ammonia per second to react with this solution until a pH of 10, to get a thick dark precipitate. 5 g of citric acid ($COOH$-CH_2-$C(COOH)(OH)$-CH_2-$COOH$) crystals dissolved in 10 ml water was added to this wet precipitate and allowed for further reaction at an elevated temperature of $80^{\circ}C$ while stirring for another 90 minutes. This sample was then washed with distilled water several times for the removal of water soluble byproducts. This is then suspended in distilled water by ultrasound treatment. The obtained fluid was kept for gravity settling of any bare nanoparticles and was then centrifuged at a rotation speed of 3500 rpm to remove any particles that may sediments [29].

Metal Particles

So far in this chapter, the discussion was on the synthesis of ferrofluids with oxide magnetic nanoparticles. However, there elementary metals like cobalt, iron, nickel etc. with huge saturation magnetisation. Hence synthesis of metal nanoparticle based ferrofluids becomes important where the application needs huge saturation magnetisation. Moreover, magnetic metal like cobalt possess huge coercivity and remamnence and hence suitable for recording like applications. However there is a big drawback of being prone to oxidation has restricted their use in most commercial applications. Hence either surface passivation is to be done before dispersing those nanoparticles in basefluid or these fluids have to be kept at an inert atmosphere. Hence shelf life of metal nanoparticle based ferrofluids is often less when compared to the oxide ferrofluids.

Metal Nanoparticle Ferrofluids from Pentacarbonyls

Metal nanoparticle based Ferrofluid can be prepared by mixing a surfactant, Sarkosyl-0 (n oleyoyl sarcosine), and metal pentacarbonyl in Decalin (decahydronaphthalene). The mixture is to be refluxed and stirred for 5—6 h at 50 C and 100 C and metal nanoparticles with average particle size of 8.5 nm can be obtained. A constant flow of argon in the reaction vessel was maintained throughout the reaction to remove CO gas [30].

Some groups reported a simple thermolysis method to synthesise metallic cobalt based nanoparticles. The synthesis involves the thermolysis of dicobalt octa-carbonyl in hydrocarbon solutions containing polymeric materials which is an easy process involving refluxing a solution of the dicobalt octa-carbonyl in toluene, until the reaction is complete, with the formation of Co metal particles with narrow size distribution [31]. However in order to make the particle size low, the presence of a polymer is needed in thermolysis process like acrylonitrile-styrene [32, 33]. In this reaction, a polymer-carbonyl complex is formed primarily which is then decomposed to form elemental cobalt.

Charles et al. and Papirer et al. showed that the particles can be surface passivated by using surfactants (diethyl-2-hexyl sodium sulphosuccinate (Manoxol-OT)) [34, 35]. These surfactants coated metal nanoparticles can be directly dispersed into the matching basefluid to obtain a stable ferrofluid.

However, the saturation magnetisation values of those metal particles are consistently lower than the saturation magnetisation of their bulk cousins [34]. This is reported to be due to the presence of carbon, spin canning and the poor crystallisation of the metals. Alloys can also be prepared by this method [36, 37].

Metal Particles by Reduction

Metal nanoparticles can be produced by the reduction of metal salts in aqueous solution using reducing agents such as sodium borohydride or sodium hypophosphite. However, mostly the particle size is very high. However nano-sized particles have been produced, the formation of which is induced by the addition of high viscous water soluble agents (like $PdCl_2$) to the solution as a nucleating agents [38-40]. These synthesised metal nanoparticles can be *insitu* surface coated with any of the surfactants of our choice so as to enable them to be dispersed in a carrier liquid to obtain a magnetic metal fluid.

CONCLUSION

The synthesis of ferrofluids can be using the age old grinding process to the novel techniques like co-precipitation, sol-gel, reverse micelle, etc. It is the properties exhibited by them that differs mainly because of the synthesis techniques. Also there is a vast choice of magnetic materials for the synthesis of these fluids. The application determines the choice of the magnetic material. For general display and other applications, magnetite is the most commonly employed material. In optical displays, magnetic materials with more transparency can be used. In biomedical applications, biocompatible maghmite is the often sought after candidate, while in transformer like applications, nickel ferrites are often chosen. Carrier fluids are also selected according to their applications. Water based ferrofluids are essential for biomedical applications. Hence an appropriate material and carrier liquid as well as compatible surfactant can be selected for our use. Synthesis techniques can also be tried so as to tune their properties to our desired levels.

REFERENCES

[1] J. Smit, H. P. Wijn J (1959) *Ferrites*. Wiley, New York.
[2] W. J. Schuele, V. D. Deetscreek (1963) *Ultrafine particles*. W. E. Kuhn (ed.) Wiley, New York.
[3] J. L. Dormann, M. Nogues *J. Phys: Condens. Mat.* 2, (1990) 1223.
[4] S. S. Papell (1965) U S Patent 3, 215, 572.
[5] R.Y. Hong, Z.Q. Ren, Y.P. Han, H.Z. Li, Y. Zheng, J. Ding, *Chem. Engg. Scie,* 62, (2007, 5912.
[6] S. E. Khalafalla, G. W. Reimers (1973) U S Patent 3,764,540.
[7] B. M. Berkovsky, V. F. Medvedev, M. S, Krakov (1993); *Magnetic Fluids, Engineering Applications*; Oxford University Press.
[8] R. E. Rosensweig (1985), *Ferrohydrodynamic*s; Cambridge University Press. : R. E. Rosenswieg; Magnetic Fluids; 124-132.

[9] J. L. Dormann, M. Nogues *J. Phys:Condens. Mat.*, 2, (1990) 1223.

[10] L. F. López,. G. Bahamón, J. Prado, J. C. Caicedo, G. Zambrano, M. E. Gómez, J. Esteve, P. Prieto, *J. Magn.Magn.Mater*, 324, (2012), 394.

[11] S. Swapna Nair, S. Rajesh, V. S. Abraham and M. R. Anantharaman, *Bull.Mater.Scie*, 34, (2011), 245.

[12] S. Swapna Nair' Francis Xavier. P.A. Joy, S. D. Kulkarni, M. R. Anantharaman, *J. Magn. Magn. Mater*, 320, 2008, 815.

[13] S. Swapna. Nair, S. Rajesh, V. S. Abraham, M.R. Anantharaman, V.P.N. Nampoori, *J. Magn. Magn. Mater*, 305, 2006, 28.

[14] V.S. Abraham, S. Swapna Nair, S. Rajesh, U.S. Sajeev and M.R. Anantharaman, Bull. Mater. Scie, 27, 2004, 155.

[15] T. J. Shinde, A. B. Gadkari and P. N. Vasambekar, J. Mater. Scie, Article available online, DOI: 10.1007/s10854-011-0474-y.

[16] T. J. Shinde, A. B. Gadkari and P. N. Vasambekar, J. Mater. Scie, 21, 2010, 120.

[17] B. Gadkari, T. J. Shinde and P. N. Vasambekar, J. Mater. Scie, 21, 2010. 96.

[18] Schuele W J, Deetscreek V D (1963), *Ultrafine particles*. Kuhn W E (ed.) Wiley, New York.

[19] K. J. Davies, S. Wells, S. W. Charles, *J. Magn. Magn. Mat.* 122: 199324-8.

[20] Nakatsuka K, Hama Y, Takahashi J Heat transfer in temperature-sensitive magnetic fluids. *J. Magn. Magn. Mat.*, 85, (1990) 207.

[21] M. George, A.M. John, S. Swapna Nair, P.A. Joy, M.R. Anantharaman, *J. Magn. Magn. Mat.*, (2006) 302 190.

[22] M. P. Pileni (1989) *Structure and Reactivity of Reverse Micelles*, Elsevier (Pub), Amsterdam.

[23] M. Gobe, K. Kon-No, A. Kitahara. *J. Coll. Int. Scie.*, 93, (1984) 293.

[24] S.W. Charles, (2002), *Ferrofluids and Magnetorheological Fluids* Ed: Stefan Odenbach, Springer Verlag, Berlin pp 3-18.

[25] R. D. K. Misra, S. Gubbala, A. Kale, W. F. Egelhoff, J.Mater.Scie. and Engg B, 111, (2004), 164.

[26] S. Palacin, P. C. Hidber, J. Bourgoin, C. Miramond, C. Fermon, G. Whitesides, *Chem. Mater.*, 8, (1996), 1316.

[27] J. P. Jolivet, R. Massart, J. M. Fruchart Nouv. *J. Chim.*, 7, (1983) 325.

[28] P. Enzel, N. B Adelman, K. J. Beckman, D.J. Campbell, A. B. Ellis, Lisensky, G.C., *J. Chem. Educ.* 1999, 76, 943.

[29] P. Reena Mary , T.N. Narayanan , V. Sunny , D. Sakthikumar , Y. Yoshida , P.A. Joy , M.R. Anantharaman , *Nanoscale Res Lett.* 5 (2010) 1706.

[30] T. W. Smith and D. Wychick, J. Phys. Chem. 84, (1980) 1621.

[31] J. R. Thomas, *J. Appl. Phys.* 37, (1966) 2914.

[32] P. H. Hess, P. H. Parker, *J. Appl. Polymer Scie.*, 10 (1966), 1915.

[33] T. W. Smith (1981) U S Patent 4,252,673.

[34] S. W. Charles, S. Wells *Magnetohydrodynamics* 26 (1990) 288.

[35] E. Papirer, P. Horny, H. Balard, R. Anthore, R. Petipas, A. Martinet *J. Coll. Int. Sci.* 94 (1983) 207.

[36] D. B. Lambrick, N. Mason, N. J. Harris, G. J. Russell, S. R. Hoon, M. Kilner *IEEE Trans. Mag.* MAG-21, (1985) 1891.

[37] D. B. Lambrick, N. Mason, S. R. Hoon, M. Kilner *J. Magn. Magn. Mat.,* 65, (1987) 257.

[38] G. Akashi, M. Fujyama (1969), US Patent 3,607,218.

[39] S. Harada, T. Yamanashi, M. Ugaji *IEEE Trans. Mag.* MAG-8: (1972) 468.

[40] S. Swapna Nair, V. Sunny, M. R. Anantharaman, *Mater. Res. Bull.* 46, (2011), 1610.

In: Ferrofluids
Editors: Franco F. Orsucci and Nicoletta Sala

ISBN: 978-1-62808-410-8
© 2013 Nova Science Publishers, Inc.

Chapter 4

OPTICAL PROPERTIES OF FERROFLUIDS: GRAIN SIZE - THE DETERMINING FACTOR

Swapna Nair[*1] *and M. R. Anantharaman*[2]
[1]CICECO, Universidade de Aveiro, Aveiro, Portugal
[2]Department of Physics,
Cochin University of Science and Technology, Cochin, India

INTRODUCTION

Twenty first century is believed to be the century of nanoscience and nanotechnology. However, the advent of nanotechnology has decades of age and a landmark of this is the famous statement by Richard Feynmann in one of his speech "There is plenty of room at the Bottom". This particular sentence has deep underlying meaning. It directly pointed towards the unraveled applications of low dimension materials. While the physicists in this century are investigating low dimension structures for understanding the causes and facts lying behind their surprising properties, materials engineers are trying to translate these superlative properties into useful applications. From the fundamental point of view, it is the particle size that determines the properties of a material. It is well known that gold in its nano size is no more a material with sparkling yellow (golden) colour. It colour changes to greenish yellow, pale green and green according to the reduction in the particle size. Hence grain size determines not only the electrical and magnetic properties, but even the appearance of a material!. The change in colour of a material directly points towards the change in the optical absorption. As we are concentrating on the synthesis and application of ferrofluids, this chapter aims at providing a detailed understanding of the tuning of the optical properties of ferrofluids by flying on the wings of nanotechnology ie, by grain size alteration and it details the electronic states, absorption processes, confinements and related phenomena that governs the optical properties of these smart fluids. Based on these observations, authors propose a method to synthesize a more transparent ferrofluid with out dilution in the magnetic properties by playing with the grain size.

---

[*] E-mail: swapna.s.nair@gmail.com.

A magnetite ferrofluid consisting of individual grains with average grain size of 12 nm will be black in colour. This is due to the colour of the magnetite nanoparticles. Actually what determines the colour of a material? Certain material is black in colour, while certain others are blue, green or yellow in colour. We know that it is the absorption (and emission process) associated with the electronic states that determines the colour of the material. That's the reason for a particular colour for a material. Hence if we want to change the colour of the material, it is necessary that we have to engineer the band structure. Here attempts are made to have a theoretical examination of the role of the grain size in tailoring the band gap and associated absorption (and hence the colour of ferrofluids). Hence authors have to introduce the basic concepts of nanotechnology in this chapter.

It is well known that nanoscience and nanotechnology is the science and technology of ultrafine particles. Particles when reduced to nanometric dimensions display novel and superlative physical, chemical, electrical, optical and magnetic properties with respect to their coarser sized cousins. At these dimensions the surface area becomes very large and the ratio of surface to bulk atoms dictates the properties of the material. Grains, grain size, intergranular interaction, quantum magnetization, quantum tunneling, superparamagnetism are some of the various phenomena exhibited by particles at the nanodimension. So manipulation of properties means the understanding of these phenomena for tailoring the properties. With the emergence of nanotechnology, newer applications of these materials are conceived and are realized as products soon.

At the same time, the real physics of many special phenomenon exhibited by the nano dimensioned materials are not clearly understood and exclusive investigations are to be carried out.

Magnetism and magnetic materials is an area where large scope for such an activity exists. Moreover nanomagnetic materials are going to play a dominant role in the day-today life of human beings. The birth of new class of materials like diluted magnetic semiconductors (DMS), spin valve transistors, giant magneto resistive (GMR) materials, ferrofluids, giant magneto caloric materials (GMC), are all indication of these. Thus study on magnetic nanocomposites and nanomagnetic materials assumes significance and is very relevant in understanding various phenomena. However their optical properties are not often investigated or analysed in depth. This chapter provides the theoretical concepts of the general optical properties of nano dimensioned materials in the beginning and the latter portion points towards the optical properties of magnetic materials and ferrofluids, which are our area of interest. The concept of making a more transparent magnetic material with grain size dependent band gap tailoring and transparency modification is discussed towards the end of this chapter.

NANOTECHNOLOGY-AN OVERVIEW

Nanotechnology is known as the technology of the 21st century which deals with the synthesis and study of ultra fine materials and their employment in technology for various applications. It can be defined as the synthesis and engineering at the molecular level for possible device applications where nanoscience deals with the investigations of phenomena and properties exhibited by materials at the nanolevel.

The demand for smaller materials for high density storage media is the fundamental motivation for the fabrication of nanoscale magnetic materials. The idea of nanoscale molecular device is not entirely new, and has been around since days immemorial. Richard Feynmann, who addressed the issue of quantum mechanical computers once opined "at any rate, it seems that the laws of physics present no barrier to reducing the size of computers until bits are the size of atoms, and quantum behaviour holds dominant sway" – a task, considering the small size that is necessary for realizing such device.

The ultimate motive in nanotechnology is to manipulate and control the individual atoms and thereby, the programmed formation of the superstructures. However, such a definition may be extended to the organization of objects having nano-dimension such as molecules/biomacromolecules and other nanoscale matters such as quantum dots, buckyballs (also known as fullerenes) and nanotubes [1, 2]

Materials consisting of particles with diameter less than 100 nm have attracted a great deal of recent research attention. Owing to their ultrafineness in size and very high surface area, these particles possess dramatic changes in physical and chemical properties as compared to their bulk counterparts which makes them ideal templates to study the physics at the nanolevel from a fundamental point of view in addition to the vast application potential in versatile fields [3, 4].

Nanoparticles behave quite different from their coarser-grained counterparts of the same composition due to the high surface to volume ratio. The more loosely bound surface atoms constitute a significant fraction of the sample and their properties influence its behaviour. For example, the melting point of gold is dramatically reduced when the particle diameter drops below 5nm. Improvements can be made in the mechanical and magnetic properties of materials.

At sufficiently small domain sizes, the particle roughness increases and becomes comparable to the radius of curvature, in the case of a spherical particle. This reduces the strength of the bonding to adjacent atoms and often results in surface atoms with increased reactivity. The abundance of particle edges and corners can create non-equivalent reactive sites on a single particle.

Their magnetic and optical properties will be profoundly modified. Smaller grained magnetic materials exist as single domain because of the energy considerations. Ferro/ferrimagnetism is reduced and below a critical diameter the materials exist as superparamagnetic. Optical properties are modified because of the quantum size effects on the band structure. Optical energy Bandgap is blue shifted for ultra fine materials. Nano sized gold is green in colour which is a semiconductor while bulk is a noble yellow metal!. We can make junctions with the same materials with different grain sizes due to the modified band structures of these ultra fine particles. This gives scope for a variety of applications in the semiconductor industry.

Particles at the Nanolevel

The physical and chemical properties of materials with their grain size approaching molecular dimensions are of great fundamental interest. Hence they became hot topic of intensive research. Nanoparticles display novel physical properties resulting from surface effects. The enhancement of the magnetization per atom and decrease in the blocking temperature from the bulk values is induced by the controlled size in a superparamagnetic nanoparticle. There has been a great deal of interest in the optical and magnetic properties of ultra fine particles in recent years.

The optical properties of materials change abruptly when the grain size becomes comparable to the Bohr radius of electron-hole pairs (excitons), quantization effects can be observed due to the quantum confinement of charge carriers in the finite volume of the particles which becomes very large when the size approaches the Bohr radius limits. Interesting and novel optical properties have been observed in the case of gold and silver when their sizes are reduced to nanodimensions.

Quantum Wells, Quantum Wires and Quantum Dots

When one of the dimensions in a crystal are made smaller so as to reach the nanoscale (a few nanometer), electrons and holes can be confined in space in along that direction as compared to their motion in a bulk crystal, which is equivalent to the reduction of one of the degree of freedom for the electrons/hole/quasi particles. These states can be termed as quantum well like states.

In quantum wires, two of the dimensions are cut off. Materials could be made into the nanowires in which one of the degrees of freedom is retained for the charge carriers electrons and holes. Hence the diameter of the nanometer is few tens of nanometers. Hence confinement is stronger as there is only one degree of freedom.

In a quantum dot, the crystal is miniaturized in all the three dimensions to a few nanometers. The greatest difference between a quantum dot as compared with a bulk crystal, quantum well and quantum wire is that Coulomb energy contribution to the ground state energy is non zero where as for quantum well and quantum wires, electrons are free and coulomb energy contribution tends to zero.

General Synthesis Techniques

There are a variety of experimental techniques for the synthesis of ultra fine magnetic particles. All these techniques need complete control over the reaction parameters.

The first technique involves the preparation of isolated particles. However ultra fine crystallites having uncontaminated free surfaces followed by a consolidation process either at room or at elevated temperatures. The specific processes used to isolate the nanostructured materials are for example, inert gas condensation, decomposition of starting chemicals and precipitation from the solutions.

1. Chemical vapour deposition (CVD), Physical vapor deposition (PVD) and some electrochemical methods have been used to deposit atoms or molecules of desired materials on suitable substrates. Nanocomposites of these materials could be synthesized in thin film form by consecutive coating of chemically different materials.
2. By introducing defects in a formerly perfect crystal another class of materials could be synthesized. This can be effected by processes like HEBM, and ion irradiation which shatters the bigger crystal.
3. The most efficient method of fine particle synthesis is the chemical co-precipitation from super saturated solutions by careful control of reaction conditions.
4. Another method being employed currently for the synthesis of metal/semi conductor oxides which is the sol-gel process in which the samples are taken in the required molar ratio and it is made into a gel and finally ignited at a higher temperature to yield nanostructured fine particles.

Of these different methods for the synthesis of nanostructured materials, inert gas condensation, physical vapour deposition, High energy ball milling and ion irradiation can be included in the physical methods for the synthesis of materials.

Chemical methods include, chemical vapour deposition, chemical bath deposition, (for thin film fabrication) chemical co-precipitation, sol-gel (for nanoparticle and composite synthesis) etc.

Synthesis of metal nanoparticles comes under a separate category as the preparation process is quite tedious with more chances of contamination so that the additional provision should be given to passivate these explodingly reactive surfaces.

Chemical Methods

Chemical methods are the most widely employed technique [5] for the synthesis of nanostructured materials owing to the versatility in design, its economic nature and its capability for tailoring the properties by carefully controlling the grain growth as it offers mixing at the nanolevel. A basic understanding of the principles of crystal chemistry, thermodynamics, phase equilibriums, phase changes and reaction kinetics is necessary to take advantage of several benefits the chemical process offer.

The grain size and size distribution, the physical properties such as crystallinity and crystal structure, and the degree of dispersion can be affected by the reaction kinetics, pH, reaction temperature, concentration of reactions, molarity and its ratio and several other factors.

There can be other factors like agglomeration of individual grains that make them unidentifiable from their bulk counter parts. There are chances of getting non stoichiometric undesired components precipitated along with the desired final product because of the slow reaction dynamics and nucleation. Hence other measures like size control, surface modification and capping are to be adopted to modify these processes.

Size Control

Control of both grain size and grain size distribution becomes important in the synthesis of nanomaterials. The grain size could be controlled by varying the synthesis parameters like concentration, choice of reactants, temperature, pH etc. Choice of templates with natural pores for the synthesis of nanoparticles is also in use. *Insitu* surface modification using stabilizers such as thiols, phosphates, and polymers prevents further grain growth. [5-10].

Grain Size Distribution

The control of grain size distribution is quite important in the synthesis of nanoparticles. To narrow the size distribution, exclusion chromatography and capillary zone electrophorosis are employed [10]. These are based on the principle of charge to size ratio is different for different grain sizes. Filtration through molecular sieves of different mesh size is also used for getting the size distribution narrower.

Size Quantization Effects

Quantization in ultra fine particles originates from the confinement of charge carriers in semiconductors with potential wells of narrow dimensions less than the De-Broglie wavelength of electron and holes. Confinements could be mere electronic, excitonic or polaronic based on the grain size and excitation energy [11]. Under these conditions, the energy bands of electrons and holes becomes close to discrete energy levels as of in atom and thus a semiconductor becomes atom like. In addition to the large change in electronic/optical properties, they also exhibit change in the effective redox potentials of photo-generated carriers.

Size quantization effects on the optical properties of semiconductors are extensively studied [12-16]. In CdS nanocrystals, a blue shift in energy band gap of 1.54 eV is obtained for a particle with radius 1nm. Blue shift is observed for many other semiconductors because of the quantum confinement effects. Trapping of charge carriers is possible for the nanostructured materials [17-22]

Optical Properties

Optical properties of ultra fine particles are profoundly modified by the grain size dependant confinement effects. In the ultra fine regime, due to very small wave function overlapping, the energy levels tends to be discrete and when the grain sizes are reduced to the order of exciton Bohr radius limit of the material, they are near molecule like materials and hence the energy levels tend to be discrete and thus there is confinement of carriers. This will alter the band gap towards the high energy limits which can be as high as 3 eV in a 5Å particle! Thus by manipulating grain sizes, materials with same chemical formulae but different band gaps can be synthesized. The influence of grain size vis a vis quantum confinement have been investigated extensively [23-25].

Nonlinear Optical Properties

Semiconductor nanoparticles are investigated because of their excellent non linear optical properties by many researchers [26-28]. In the linear regime, photoinduced blue shift in absorption edge is observed, which causes transient bleaching in nonlinear regime [29]. Third harmonic generation, and free carrier absorption are also observed in semiconductor nanoparticles especially in CdS nanoclusters [30]. Due to near molecule like energy levels, they show strong Saturable Absorption (SA), Reverse Saturable Absorption (RSA) and multi photon absorption.

Towards Transparent Magnetic Materials: By Grain Size Reduction

Transparent magnetic materials finds extensive application potential in versatile fields like Xerox technology, magneto-optical recording, magnetic field controlled optical modulators, magneto-optical displays and switching devices [31-33]. Aerogel based transparent magnetic materials have been proposed now a days which is rather too weekly magnetic to be used for any applications [34].

With the advent of nanotechnology, it is known that both the optical and magnetic properties could be greatly modified if the synthesized particles are in ultra fine regime. The optical properties of the materials change dramatically when the grain size is reduced to the order of Bohr radius limit. Semiconductor nanocrystallites are recently studied for their excellent optical properties due to the quantum confinement effects. It is reported first in CdSe nanocrystals that in the strong quantum dot confinement a shift of 1.54 eV in energy band gap is exhibited [35]. Also the enhanced blue shift due to weak exciton confinement is reported in γ Fe_2O_3 nanocrystals [36]. A strong and dot like confinement can occur if the grain size could be further reduced below Bohr radius of the material which can enhance the quantum confinement effects to a higher order.

Magnetic properties also change with the ultra fine nature of the grains. Normally materials exhibit single domain and superparamagnetic characteristics at the nanolevel. However if the grain size is reduced further, thermal energy contribution becomes higher which can randomize the spins giving negligible value of magnetization and thus becomes non favorable from the application point of view.

Alloying with any other metals is also found to enhance the blue shift in Ni doped CdTe nanocrystalline thin films [37]. However such a technique is never found to be employed in enhancing the optical transparencies of a magnetic material. Ferrofluids, are stable colloidal suspension of ultra fine magnetic particles in a suitable base fluid [38], are ideal candidates to study the magnetism and optics at the nanolevel. However, due to the reduction in magnetization with reduction in grain size of the order of 40 Å, alternative methods have to be employed to enhance the optical transparencies other than the grain size reduction.

Studies on the size effects or alloying effects on the optical properties possess problems and careful methods of preparation have been employed so as to prepare magnetic fluids having high saturation magnetization values with least possible grain size (50-65Å) and a narrower size distribution.

Semiconductor Nanocrystals

It has been very helpful for the Physics of nanocrystals to employ the simplified models to trace the effects originating from the three dimensional spatial confinements. An extension of the effective mass approximation towards the spatially confined structures leads to a particle in a box problem and provides a way to calculate the properties of nanocrystals that are not possible to analyze by other means. Hence in nanocrystals, "crystal to nanocluster approximation" and "single molecule to quantum dot" approximations are examined for their respective employment in our system with known particle diameters.

Excitons

Excitons are quasiparticles formed by the elementary excitations of the system consisting of real particles[39]. By this approach electrons in the conduction band can be considered as the primary elementary excitation of the electrons in the crystal. This will lead to a quasi particle *hole* in the valence band with a proper quasi momentum providing a quasi effective mass. Hence the ground state of an electron is a vacuum state and the first excited state is the creation of one electron-hole pair. Electrons and holes can be described by Fermi Dirac statistics. Ideal excitation produces the electrons and holes with no attraction. However there is a coulomb attraction for electrons and holes which results in the formation of a quasi particle called an exciton. Exciton possess a Bohr radius of

$$a_B = \frac{\varepsilon \hbar^2}{\mu e^2} = \varepsilon \frac{m_0}{\mu} 0.53 \overset{\circ}{A}$$

As the reduced electron mass is smaller than electron mass and dielectric constant is larger several times than in vacuum, the excitons possess large Bohr radius as compared to the ground state of hydrogen atom although they hydrogen like states.

Quantum Confinements

In semiconductors, the de Broglie wavelength of an electron and a hole and the Bohr radius of an exciton may be considerably larger than the lattice along the confinement direction. For d=0, complete confinement occurs with zero dimensional structure which differs from quantum wells and wires and the density of states is a discrete delta function resulting in the finite motion in all three dimensions with finite number of atoms and elementary excitations in each quantum dot. This gives strong confinements.

Weak and Strong Exciton Confinements

Weak confinement regime corresponds to a region when the particle radius is small, but a few times greater than exciton Bohr radius. In this quantization of exciton centre of mass occurs. In strong confinement, in contrary to the hydrogen like Hamiltonian, the appearance of potential does not allow the centre of mass motion and the motion of a particle with reduced mass to be considered independently. Strong confinements are resulted only when the grain size is several times less than the exciton Bohr radius and this grain size distribution is quite narrow.

A Heavy Hole and a Light Hole

A heavy hole and a light hole can be differentiated by their energy curves. Light holes have a sharp energy band and the energy parabola by the first approximation will be steeper, while the heavy hole has a broader energy curve. Their density of states is very high in normal crystals and there by influencing all the optical properties exhibited by these crystals. Light holes are lesser in number and their density of states are thus very low and unless the grains fall in dot regime, ie, when the grain size becomes much below the Bohr radius limit of excitons, the properties of the crystal wont be much affected by their presence. However, in the highly confined regime, light holes also results in the formations of excitons with a very small effective mass for the hole, there by changing the energy shift towards the blue region. Electron-Heavy hole pair exciton has almost the same effective mass in all semiconductors, while the excitons formed by a light hole and electron has varying effective mass in different crystals.

Generally if there is a triplet and singlet level, triplet states provide high momentum and hence higher effective mass to the exciton, thus the hole is a heavy hole, while the singlet state gives rise to a light hole.

Optical absorption spectrum of many of the nanocrystalline semiconductors exhibits a blue shift due to the quantum confinements [40-43]. The confinements in these nanostructured semiconductors could be broadly divided into two extremes, namely the strong and the weak confinement. In the strong confinement regime, the grain size is less than $2a_0$ where a_0 is the exciton Bohr radius of the material and in the weak confinement regime, the grain size is larger than $4a_0$. In between these limiting cases, both electron and hole confinement and their Coulomb interaction should also be considered.

Strong quantum confinement will occur only if the grain size distribution is very narrow and the particle could be compared as a molecule in its orbital level in which a quantum dot like confinement occurs if the nanocrystallite grain size is less than the Bohr radius of the material.

In the weak confinement regime, quantisation of exciton centre of mass comes into play. Starting from the dispersion law of an exciton in a crystal, the energy of a free exciton is replaced by a solution derived for a particle in a spherical potential well [44]. The energy of an exciton in the weak confinement case is then of the form

$$E_{nml} = E_g - \frac{Ry^*}{n^2} + \frac{\hbar^2 \chi_{ml}^2}{2Ma^2} \tag{1}$$

E_g is the bandgap, Ry^* is the Rydberg's constant.
For the lowest state (n=1, m=1, l=0) the energy can be expressed as

$$E_{1s1s} = E_g - Ry^* + \frac{\pi^2 \hbar^2}{2Ma^2} \tag{2}$$

which could be rewritten as,

$$E_{1s1s} = E_g - Ry^* \left[1 - \frac{\mu}{M} \left(\frac{\pi^2 a_B^2}{a} \right) \right] \tag{3}$$

where μ is the reduced mass of the electron-hole pair. The value $\chi_{10} = \pi$, and the relation for confinement energy becomes,

$$\Delta E_{1s1s} = \frac{\mu}{M} \left(\frac{\pi a_B}{a} \right)^2 Ry^* \tag{4}$$

which is, however, small compared with Ry* so far as $a >> a_B$ holds. This is the quantitative justification of the term "weak confinement".

Taking into account of photon absorption which can create an exciton with zero angular momentum, the absorption spectrum will then consist of a number of lines corresponding to states with l=0. therefore, the absorption spectrum can be derived from eqn 9.3 with $\chi_{m0} = \pi m$ radius, using eqn 3,4 and 5, confinement energy can be written as

$$E_{sh} = \frac{h^2 \pi^2}{2m_{eff} R^2}, \tag{5}$$

where R is the nanocrystalline radius and m_{eff} is the effective mass of the exciton in weak confinement. Thus confinement energy is clearly a function of the nanoparticle radius and varies as $1/R^2$.

The exciton of 1s electron-heavy hole has comparable values of effective mass m_{eff} for almost all the semiconductors and thus the energy shift variation becomes a function of the nanoparticle radius in the weak confinement regime.

Now come back to ferrofluids. We can concentrate primarily the most widely investigated ferrofluids, ie, those based on magnetite as the magnetic material. The colour of such ferrofluids are black.

How to make the transparency of that black ferrofluid? Generally the first trend is to mix it in a transparent matrix or to go for dilution of the ferrofluid with a transparent carrier liquid. But imagine what will be the result. We have to sacrifice greatly the magnetic properties of these fluids. We will have to forget the applications of these fluids in sealing, speaker coil, magnetic displays etc.

Then what is the next alternative to enhance the optical properties of these fluids?

After reading about excitons, quantum confinement and grain size dependent band gap shift in the former half of the chapter, readers can quickly suggest the second alternative for the enhancement of transparency of these fluids.

It is tailoring the grain size for tuning the band gap. Generally the ferrofluids are composed of magnetic nanoparticles with average grain size of 10 nm. If the grain size is reduced to the order of 7nm, the shift in band gap from the bulk is almost double when compared to the fluid comprising of nanoparticles with average grain size of 10 nm!!

This means that we can slowly shift the absorption from IR range to Visible by playing with the grain size which is quite important form the application point of view especially regarding the magneto optical and optical applications of these smart fluids. However, we cannot indefinitely reduce the diameter of the nanoparticles as it will make the nanoparticles more and more paramagnetic which is not desirable from the application point of view. Hence there will be a trade off between the magnetism and transparency and the average diameter if we are playing with only the grain size.

Now consider eqn. 5. We have another option to change the band gap other than by changing the diameter of the particles. It is to play with the second factor m_{eff}.

How to change the effective mass of the exciton pair?

The answer is DOPING/SUBSTITUTION!!. Other metals like nickel, manganese etc. can be substituted in place of Fe^{2+} which have a different heavy hole energy spectra and associated change in the orbital hybridization and spin momenta and hence a change in effective mass can be obtained with their substitution which can result in a further band gap shift with out affecting the magnetic properties to a great extent.

CONCLUSION

Transparent magnetic materials are the dream of scientists and technologists working in the area of magneto optics. Enhancing transparency is often done by the dilution of the magnetic materials in a transparent host matrix which will result in the dilution of magnetic properties. Here an alternative method is suggested for the enhancement of the transparency of ferrofluids by reducing the particle radius of the magnetic material suspended in the fluid towards the Bohr radius limit. The possibility of changing the band gap by doping/ substitution is also discussed. Necessary theoretical concepts are also provided. Hence a proposition is made for the synthesis of a more transparent ferrofluids by tailoring the grain size is provide in this chapter which can be quite revolutionary if realized experimentally.

REFERENCES

[1] J. Chatterjee, Y. Haik, C.J. Chen, *J. Magn Magn. Mater*. 246 (2002), 382.

[2] Q. A. Pankhurst, J. Connolly, S K Jones and J Dobson, *J. Phys. D:Appl.Phys*.36 (2003) R167.

[3] M. Shinkai, M. Yanase, M. Suzuki, H. Honda, T. Wakabayashi, J. Yoshida, T. Kobayashi, *J. Magn Magn. Mater*.194, (1999), 176.

[4] Urs O Hafeli, Gayle J. Pauer, *J.Magm Mag. Mater*. 194 (1999),76.

[5] T. Rajh, O.I Micic, A.J Nozik, *J. Phys.Chem*, 97, (1993) 11999.

[6] Y.Nosaka, K. Yamaguchi, H.Miyama, H. Hayashi, *Chem.Lett*, (1988) 605.

[7] S. Mahamuni, A.A Khosravi,M. Kundu, A Khirsagar, A Bedekar, D.B Avasare, P. Singh, S. K Kulakarni, *J. Appl. Phys*, 73, (1993) 5237.

[8] H. Matsumoto, T. Sakata, H. Mori and H. Yoneyama, *J. Phys. Chem*, 100, (1996)13781.

[9] C. Murray, D. Norris, B.Bawendi, *J. Am. Chem. Soc*, 115, (1993) 8706.

[10] H. Matsumoto, H. Uchida, T. Matsunaga, K. Tanaka, T. Sakata, *J. Phys.Chem,* 98, (1994) 11549.

[11] Harisingh Nalwa, (2002) *Nanostructured materials and nanotechnology*, Academic Press London.

[12] Henglein, *Chem Rev* 89,(1989) 1861.

[13] H. Weller, H.M Schmidt, U. Koch, A. Fojtik, S. Baral, A Henglein, W. Kunath, K. Weiss, E. Diemen, *Chem. Phys. Lett*, 124, (1986) 557.

[14] H. Weller, *Adv.Mater*, 5, (1993) 88.

[15] D. Heyes, O.I Micic, T. Nenadovic, V. Swayambhunathan, D. Miesel, *J. Phys. Chem*, 93, (1989) 4603.

[16] C. H Fischer, A. Henglein, *J. Phys*. 93, (1989) 5578.

[17] R.V. Kamat and B. Patrick, *J.Phys.Chem*, 96, (1992) 6829.

[18] J. Bedja, S. Hotchandani, P. V. Kamat, *J. Phys.Chem*, 97, (1993) 11064.

[19] J. Bedja, S Hotchandani, P.V kamat, *J. Phys.Chem* 98, (1994) 4133.

[20] D. E Skinner, D. P. J Colombo, J. J. Caveleri, R.M. Bowman, *J.Phys.Chem*, 99, (1995) 7853.

[21] J. G. Zhang, R.H.O. Neil, T. W. Robertie, *J.Phys.Chem*, 98, (1994) 3859.

[22] J. Z. Zhang, R. H. O. Neil, T. W Roberti, J. L. McGowan, J. E. Evans, *Chem.Phys.Lett*, 218, (1994) 479.

[23] C. Allan, M. Delerue, M. Lannoo, *Appl. Phys. Lett*, 70, (1997) 2437.

[24] M. Goryll, L Vescan, K. Schmidt, S. Mesters, H. Luth, and K Szot *Appl. Phys.Lett*, 71 (1997) 410.

[25] G. Allan, C. Delerue, M. Lannoo *Appl. Phys. Lett*, 71 (1997), 1189.

[26] L.T Cheng, N.Herron, Y.Wang, *J.Appl.Phys*, 66, (1989) 3417.

[27] C.Y Yeh, S.B Zhang, A Zunger, *Appl.Phys.Lett*, 64, (1994) 3545.

[28] A.J Heliweil, R.M Hochstrasser, *J.Chem.Phys*, 82, (1985) 179.

[29] E.F Hilinski, P.A Luca, Y Wang, *J. Chem. Phys*, 89, (1988) 3435.

[30] L.T Cheng, N.Herron, Y.Wang, *J.Appl.Phys*, 66, (1989) 3417.

[31] M. Zayat, F. D. Monte, M. P. Morales, G. Rosa, H. Guerrero, Carlos J. Serna and D. Levy *Adv.Mater.* 15 (2003) 1809.

[32] M. Gich *et al.* 2003 *Appl.Phys.Lett* 82 4307.

[33] Z. H. Zhou, J. M. Xue, H. S. O. Chan, J. Wang, *J.Appl.phys.* 90 (2001) 4169.

[34] F.D. Monte, M. P. Morales, D. Levy, A. Fernandez, M. Ocana, A. Roig, E. Molins, K. O. Grady , C. J. Serna *Langmuir* 13 (1997) 3627.

[35] George C Hadjipanayis and Richard W Siegel, (2002) *Nanophase materials-synthesis, properties and applications*, kluwer Academic Publishers, London, 432.

[36] S. Swapna Nair, Mercy Mathews, M.R. Anantharaman, *Chem. Phys. Lett*, 406, (2005) 283-526.

[37] O. Alvarez-Fregoso, J.G. Mendoza- Alvarez, and O. Zelaya-Angel, *J. Appl. Phys*, 82 (1997), 708.

[38] Ronald .E. Rosenweig (1985), *Ferrohydrodynamics*, Cambridge University Press, London.

[39] Richard Turton, (2000) *The Physics of Solids*, Oxford University Press, London.

[40] D. E. Skinner, D.P.J Colombo, J.J Caveleri, R.M Bowman, *J.Phys.Chem*, 99, (1995) 7853.

[41] J. G. Zhang, R.H.O' Neil, T.W Robertie, *J.Phys.Chem*, 98, (1994) 3859.

[42] J. Z. Zhang, R.H.O' Neil, T.W Roberti, J.L McGowan, J.E Evans, *Chem. Phys. Lett*, 218, (1994) 479.

[43] M. Xu and P. J. Ridler, *J. Appl. Phys.* 82 (1997), 326.

[44] Michael Quinten, *Optical properties of nanoparticle systems*, John Wiley and Sons, 2010, Eds: Mie and Beyond. pp 154.

In: Ferrofluids
Editors: Franco F. Orsucci and Nicoletta Sala

ISBN: 978-1-62808-410-8
© 2013 Nova Science Publishers, Inc.

Chapter 5

NON LINEAR OPTICAL PROPERTIES OF FERROFLUIDS INVESTIGATED BY Z SCAN TECHNIQUE

Swapna Nair[1,], Jinto Thomas[2,3], Suchand Sandeep[2], M. R. Anantharaman[4] and Reji Philip[2]*

[1]Departmento de Eng.Cerâmic, Vidro and CICECO, Universidade de Aveiro,
Aveiro, Portugal
[2]Light and Matter Physics Group, Raman Research Institute, Bangalore, India
[3]Institute for Plasma Research, Gandhinagar, India
[4]Department of Physics, Cochin University of Science and Technology Kochi, India

ABSTRACT

This chapter gives a general outline of the optical nonlinearity shown by materials, and it also deals with the nonlinear optical properties of ferrofluids measured by open aperture Z-scan technique. Magnetite based ferrofluids were investigated for their non linear properties like induced scattering, multiphoton absorption etc and the observation were linked with the general non linear optical theories and the data were fitted for multiphoton absorption coefficients. The chapter gives the results of studies conducted on the non linear optical properties of ferrofluids.

Surfactant coated ferrofluids based on magnetite and nickel doped magnetite have been synthesized by the co-precipitation process, with oleic acid as the surfactant and kerosene as the carrier. The colloidal suspensions contain nanosized particles of approximately 80 Å diameter, with a number density of the order of $10^{22}/m^3$.

Optical nonlinearities associated with the samples have been studied in the nanosecond and femtosecond excitation domains, employing the open aperture Z-scan technique. Excited state multi photon absorption and induced scattering are found to be responsible for the observed optical limiting in the nanosecond domain, while electronic nonlinearities dominate in the femtosecond regime, which were fitted for three photon

---

[*] E-mail: swapna.s.nair@gmail.com.

absorption. The wavelength and concentration dependence of the limiting also is investigated.

INTRODUCTION

Ferrofluids are stable colloidal suspensions of nanomagnetic materials suspended in a suitable base fluid [1]. These rheological fluids have attracted the attention of scientists and engineers due to their possible applications in versatile fields. Surfactant separated ferrofluids assume significance due to their ease of preparation, and greater stability against agglomeration. This stability is achieved by coating the fine magnetic particles with molecules like oleic acid, which have a polar head, thereby ensuring that the ultrafine particles are separated. Thus a stable colloidal suspension is obtained, with the suspended fine particles executing constant Brownian motion [2].

Magnetite ferrofluids are the oldest and widely used ferrofluids due to their very high saturation magnetization, which finds enormous applications in a variety of fields [3]. Due to their good thermal stability and stability against agglomeration, these smart fluids have been extensively used in many engineering applications such as in loud speaker coils and pressure sensors. Fluid dynamics of ferrofluids in the presence of an applied magnetic field is governed by the modified Bernoulli's equation, given by

$$1/2\rho v^2 + \rho gh + P - \mu_0 \int MdH = constant$$

where the additional term arises due to the interaction of individual magnetic moments with the applied magnetic field [1]. In the absence of an external magnetic field, the equation will be reduced to the standard Bernoulli's equation. The magnetic field induced structural anisotropy of ferrofluids leads to many special magneto-optical properties like field induced optical birefringence, linear and circular dichroism, Faraday rotation and ellipticity. They also show many field independent properties like zero field birefringence [4-8].

Ever since powerful laser systems became commercially available, there has been a growing need for the protection of human eyes and optical sensors from intense laser radiation. The development of optical limiters has received significant attention in this regard. An ideal optical limiter should be transparent to low energy laser pulses, and opaque at high energies. Several organic and inorganic compounds have been found to exhibit good optical limiting properties [9-15]. Nanomaterials such as metal and semiconductor nanoparticles also have been investigated in much detail [16-22]. However, the same has not been reported for ferrofluids so far. In fact, it is the physical and chemical stability of ferrofluids, which is an important attribute for an optical limiter, which prompted us to investigate their optical limiting properties.

We employed the technique of Z-scan, which is a simple and highly sensitive method widely used for the determination of nonlinear optical constants of materials. The present samples have been irradiated using 7 ns (532 nm), and 100 fs pulses (400 nm and 800 nm). The wavelength, concentration and pulse width dependence of the limiting process is investigated.

The Z- Scan Experiment

Characterization of the nonlinear optical properties of materials is of utmost interest in several fields of physics, both from the fundamental and the applied points of view. In particular, great effort has been devoted to the determination of the third-order nonlinear optical susceptibility, $\chi^{(3)}$, responsible for phenomena such as third harmonic generation or optical phase conjugation. In media with inversion symmetry (such as gases, liquids and non-crystalline materials), third order nonlinearity is the lowest order nonlinearity allowed under the electric-dipole approximation.

Third harmonic generation, phase conjugation, saturation, self-focusing, optical Kerr effect, and two-photon absorption can all be attributed to this optical nonlinearity. There has been a great deal of research directed to investigating these phenomena in various materials and in pursuing their application.

The nonlinear intensity index of refraction n_2 and the two-photon absorption coefficient β are normally used to quantitatively characterize these kinds of self-action nonlinear optical behavior. Many experimental techniques have been developed to measure the magnitude and dynamics of third order nonlinearities.

Two most commonly used methods are Degenerate Four Wave Mixing (DFWM) and beam distortion measurements (Z-scan). DFWM directly measures the third order nonlinear susceptibility and usually involve a complicated experimental set up. The Z-scan method utilizes the self-focusing effects of the propagation beam to measure the nonlinear refractive index. The most commonly used method to extract β, is the nonlinear transmittance method which measures the transmittance of the laser medium as a function of the laser's intensity.

Among the many methods of measuring $\chi^{(3)}$, the Z-scan technique is the simplest and deserves the most attention. As stated above, the Z-scan technique is a simple and popular experimental technique to measure intensity dependent nonlinear susceptibilities of materials. Moreover, it is a highly sensitive single beam technique for measuring both non-linear refractive index and nonlinear absorption coefficients for a wide variety of materials. In this method, the sample is translated in the Z-direction along the axis of a focused Gaussian beam, and the far field intensity is measured as function of sample position.

Analysis of the intensity versus sample position Z-scan curve, predicated on a local response, gives the real and imaginary parts of the third order susceptibility. In this technique the optical effects can be measured by translating a sample in and out of the focal region of an incident laser beam. Consequent increase and decrease in the maximum intensity incident on the sample produce wavefront distortions created by nonlinear optical effects in the sample being observed.

Moving the sample along a well defined focused laser beam, and thereby varying the light intensity in the sample one obtains the z-scan. By varying the size of an aperture kept in front of the detector, one makes the z-scan transmittance more sensitive or less sensitive to either the real or imaginary parts of the nonlinear response of the material, i.e., nonlinear refractive index and nonlinear absorption, respectively. The z-scan method is an experimental way to obtain information on nonlinear refractive index and nonlinear absorption properties of materials. With nonlinearity, here we simply mean the intensity dependent response of the material, which can be used to obtain an optical limiting device, either by nonlinear refraction or by nonlinear absorption such as TPA or three photon absorption.

The other widely used technique, degenerate four-wave mixing (DFWM), involves a far more complex experimental apparatus but provides several advantages. The fact that the setup includes temporal and spatial overlapping of three separate beams permits increased flexibility, such as the possibility of measuring different tensor components of $\chi^{(3)}$, and a straight forward study of temporal behavior.

When the Z scan was originally devised, it was used to characterize the nonlinear susceptibility of transparent bulk materials. Nevertheless, its use has now been extended to the study of a wide variety of samples.

In particular, it is often used to study absorbing media. Among the latter, its use for the assessment of materials consisting of semiconductor or metal crystallites of nanometer size embedded in dielectric matrices has been very common. These composites are currently the object of intensive research, their most direct application being related to their high third-order nonlinear susceptibility, $\chi^{(3)}$, which makes them promising candidates for the development of all-optical switching devices.

Z-scan is based on the principles of spatial beam distortion and offers simplicity as well as very high sensitivity. There are two types of Z-scan techniques, namely, the 'Closed aperture' Z-scan and the 'Open aperture' Z-scan.

The Z-Scan Experimental Setup

In a typical experimental setup, a lens initially focuses a laser beam with a transverse Gaussian profile. The sample, the thickness of which is kept less than the Rayleigh range, is then moved along the axial direction of the focused beam in such a way that it moves away from the lens, passing through the focal point. At the focal point, the sample experiences the maximum pump intensity, which will progressively decrease in either direction of motion from the focus.

A suitable light detector is placed in the far field and the transmitted intensity is measured as a function of the position of the sample, to obtain the open aperture Z-scan curve. Then an aperture of suitable S value is placed closely in front of the detector, and the experiment is repeated to obtain the closed aperture Z-scan.

The absorptive nonlinearity is first determined from the open aperture data, and then the refractive nonlinearity is determined from the closed aperture data.

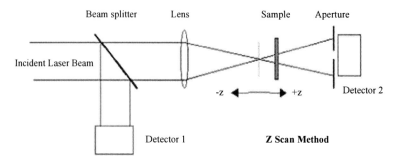

Figure 1. Schematic of Z scan experimental set-up.

Figure 2. A Photograph of the Z-scan experimental setup in the nanosecond laser pulses.

Closed Aperture Z-Scan

In this method, a single Gaussian beam in tight focus geometry is allowed to pass through a non-linear medium. The sample is moved along the z-direction so that it passes through the focal region of the beam. The transmitted beam is passed through an aperture placed in the far field, and then measured by the detector, for different values of sample position z. This method gives quantitative information on the non-linear refraction of the sample.

Assume, for example, a material with a negative non-linear refractive index and a thickness smaller than the diffraction length of the focused beam (a thin medium). This can be regarded as a thin lens of variable focal length. Starting the sample scan from a distance far away from the focus and close to the lens (negative z), the beam irradiance is low and negligible non-linear refraction occurs. Hence the transmittance (D_2/D_1) remains relatively constant. As the sample is brought closer to the focus, the beam irradiance increases, leading to self-lensing in the sample (As the sample is moved towards the focus, the intensity increase, since the width of the beam decreases towards the focus and the entire energy concentrates into a small area.). A negative self-lensing prior to focus will tend to collimate the beam, causing a beam narrowing at the aperture, which results in an increase in the measured transmittance. When the sample passes the focal plane to positive z, the same self-defocusing results in a more diverged beam at the aperture, causing a decrease in transmittance. This suggests that there is a null as the sample crosses the focal plane. This is analogous to placing a thin lens at or near the focus, resulting in a minimal change of the far field pattern of the beam. The Z-scan is completed as the sample is moved away from the focal plane to the far field in the positive z direction. In the far field since the beam radius is large, the irradiance is again low so that only linear effects will be present.

A prefocal transmittance maximum (peak) followed by a post focal transmittance minimum (valley) is therefore, the Z-scan signature of negative refractive index nonlinearity.

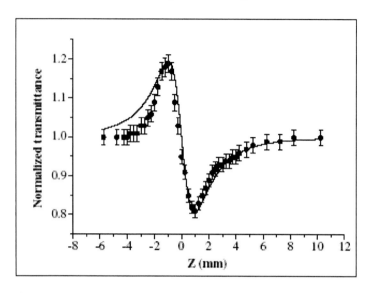

Figure 3. A typical closed aperture z- scan curve.

Positive nonlinear refraction, following the same analogy, gives rise to the opposite valley-peak configuration. It should be noted here that the sensitivity to nonlinear refraction is due to the presence of the aperture, and removal of the aperture will completely eliminate the effect. Thus an extremely useful feature of the Z-scan method is that the sign of the nonlinear index is immediately obvious from the data. Furthermore, its magnitude can also be readily estimated using a simple analysis for a thin medium. For a transparent nonlinear medium, the Z-scan curve will be symmetrical about the z = 0 point. However, if absorptive non-linearities are present, the symmetry will be lost. For example, multiphoton absorption will suppress the peak and enhance the valley, while saturable absorption leads to the opposite effect.

For a given $\Delta\phi_0$, the magnitude and shape of T(z) do not depend on the wavelength or geometry as long as the far field condition for the aperture plane ($d >> z_0$) is satisfied. The aperture size S, is an important parameter since a large aperture reduces the variations in T(z). This reduction is more prominent in the peak where beam narrowing occurs and can result in a peak transmittance, which cannot exceed (1- S). Thus for very large aperture or no aperture (S=1), the effect vanishes and T(z) = 1 for all z and $\Delta\phi_0$. For small $|\Delta\phi_0|$, the peak and valley occur at the same distance with respect to focus, and for a cubic nonlinearity, the distance is found to be 0.86 z_0. With larger phase distortions ($|\Delta\phi_0| > 1$), numerical evaluation of equations shows that this symmetry no longer holds and peak and valley move towards ± z for the corresponding sign of nonlinearity($\pm\phi_0$) such that their separation remains nearly constant given by,

$$\Delta Z (p, v) \cong 1.7\, z_0 \tag{1}$$

There is another easily measurable quantity, $\Delta T_{(p,v)}$, is the difference between the normalized peak and valley transmittance: $T_p - T_v$. For a given order of nonlinearity, the curves

$\Delta T_{(p,v)}$ versus $|\Delta\phi_0|$ exhibits universality, i.e., they are independent if laser wavelength, geometry(as long as the far field condition is met) and the sign of nonlinearity. For all aperture sizes, the variation of $\Delta T_{(p,v)}$ is found almost linearly dependent on $|\Delta\phi_0|$ as given by the equation,

$$\Delta T_{(p,v)} \approx 0.406 \, (1-S)^{0.25} \, |\Delta\phi_0|, \text{ for } |\Delta\phi_0| \leq \pi \qquad (2)$$

which can be used to estimate the non-linear refractive index. If the experimental apparatus and the data acquisition system are capable of resolving transmittance changes of $\Delta T_{(p,v)} \cong 1$ %, then a phase change corresponding to a wavefront distortion as low as $\lambda/250$ can be measured. Achieving such sensitivity, however, requires relatively good optical quality of the sample under study.

These equations were obtained based on a cubic nonlinearity (i.e. a $\chi^{(3)}$ effect). A similar analysis can be performed for higher order nonlinearities. Regardless of the order of nonlinearity, the same qualitative features are to be expected from the Z-scan analysis. For a fifth order effect, assuming a thin sample and using the GD approach, the peak and valley are found to be separated by $\cong 1.2 \, z_0$ as compared to $1.7 \, z_0$ obtained for the third order effect.

Open Aperture Z-Scan

Large refractive nonlinearities in materials are commonly associated with a resonant transition, which may be of single, or multiphoton nature. The non-linear absorption in such materials arising from either direct multiphoton absorption, saturation of the single photon absorption, or dynamic free carrier absorption have strong effects on the measurements of non linear refraction using the Z-scan technique. Clearly, even with nonlinear absorption, a Z-scan with a fully opened aperture (S=1) is insensitive to non-linear refraction.

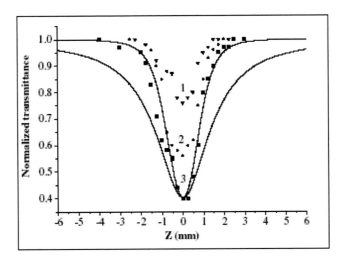

Figure 4. Open aperture z- scan at different intensities ((1). 2.2×10^{11}W/cm^2, (2).3.2×10^{11}W/cm^{-2}., (3).4×10^{11}W/cm^{-2}.). Dashed line is a theoretical fit for two photon absorption and solid line is a theoretical fit for the three-photon absorption at the intensity of 4×10^{11}W/cm^{-2}.

In the Open Aperture Z-scan Technique, the sample transmission for different z values is measured as before, but the aperture is removed from the set up. The absorptive non-linearity will be a maximum at the focal plane. For example, in the case of two-photon absorption, the transmittance will be a minimum and in the case of saturable absorption, the transmittance will be a maximum. The open aperture Z-scan curve is symmetrical about the focus (z=0).

Although this method has several advantages, disadvantages of this technique include the fact that it requires a high quality Gaussian TEM_{00} beams for absolute measurement. The analysis must be different if beam is non-Gaussian. Sample distortions or wedges, or a tilting of the sample during translation can cause the beam to walk off the far field aperture. This produces unwanted fluctuations in the detected signal. Even if these are kept under control, beam energy jitter will produce the same effect. A second reference arm may be employed to subtract out the effects of beam jitter.

The Regime of Linear Optics

Nonlinear optics is a study that deals mainly with various new optical effects and novel phenomena arising from interactions of intense coherent optical radiation with matter. Nonlinear optical effects could not be experimentally observed in the pre-laser era, since the field strengths of conventional sources have been much too small to perturb the atomic and inter atomic fields. Naturally, light waves with low intensities are not able to affect atomic fields to the extent of changing optical parameters. Thus the assumption of linearity of an optical medium has the following far reaching consequences: the optical properties, such as the refractive index and the absorption coefficients are independent of the light intensity; the frequency of light cannot be altered by its passage through the medium; and light cannot interact with light; that is, two beams of light in the same region of a linear optical medium have no effect on each other. The nonlinear optical properties of materials play a crucial role in today's technology .For example, optical switching which is necessary for optical computing maybe realizable via the non-linear processes of materials. Another example is optical storage, where laser light is used as a means of reading or writing information. If the wavelength of the laser light is halved the storage capacity of the device will quadruple. This may be achieved by frequency doubling the fundamental wavelength in a nonlinear material through nonlinear second harmonic generation. Clearly, the nonlinearity of materials enhances our optical capability and allows us to improve our optical devices [23-26].

The behavior of a dielectric medium through which an electromagnetic (optical) wave propagates is completely described by the relation between the polarization density vector P($\mathbf{r}$,t) and the electric field vector E($\mathbf{r}$,t).A transparent dielectric medium placed in an electric field becomes electrically polarized, and the displacement of the electron density away from the nucleus results in a charge separation (an induced dipole) with dipole moment P_i .With small applied fields the linear response approximation holds, so that the displacement of charge from equilibrium position is proportional to the strength of the field;

$$P_i = \alpha' E \tag{3}$$

where α' is the linear polarizability of the molecule or atom, and E is the applied electric field. If the field oscillates with a frequency then the induced polarization will have the same frequency and phase if the response is instantaneous. Also, the dipole moment vector per unit volume (i.e. the polarization density) P is given by $P = \Sigma P_i$, where the summation is over the dipoles in unit volume. For a linear dielectric medium,

$$P = \varepsilon_0 \chi E \tag{4}$$

where ε_0 is the permittivity of free space and χ is the dielectric susceptibility of the medium. The quantity χ is a constant only in the sense of being independent of E; its magnitude is a function of frequency.

Nonlinear Optics

However, after the advent of lasers, the coherence of the laser beam enabled an examination of the behavior of light in optical materials at intensities higher than that was previously possible. Irradiation of a medium with high intensity laser radiation is, in principle equivalent to the application of a large electric field to the material. Many experiments revealed that the optical media do in fact exhibit nonlinear behavior. Under such circumstances the following phenomena takes place

The refractive index consequently the speed of light in an optical medium does change with intensity

1. Light can alter its frequency as it passes through a nonlinear optical material
2. Light can control light; photons can be made to interact.

Thus, when a molecule is subjected to strong laser radiation, the molecule's polarization is being driven beyond the linear regime. The modified nonlinear polarization P_m (which is a function of the applied field and leads to nonlinear effects) is expressed as

$$P_m = \alpha' E + \beta' E^2 + \gamma' E^3 + \tag{5}$$

where,

α' = linear polarizability of the molecule or atom
β' = first molecular hyperpolarizability (second order nonlinearity term)
γ' = third order nonlinearity term
E = electric field acting on the molecule

With increasing field strength, nonlinear effects become observable due to the presence of the higher powers of E in eqn.(5). (α is usually much greater than β and γ) Eqn.(4) also gets generalized to

$$P_m=\varepsilon_0(\chi^{(1)}E+\chi^{(2)}E^2+\chi^{(3)}E^3+\ldots\ldots\ldots) \tag{6}$$

where $\chi^{(1)}$ is the first order susceptibility ,and $\chi^{(n)}$ is the nth order nonlinear optical susceptibility. Thus optical characteristics of a medium such as dielectric permittivity, refractive index, etc, which depend upon susceptibility, also become functions of the field strength E. It may be noted here that optical nonlinearity is a property of the medium through which light travels, rather than a property of light itself.

Suppose that the field incident on a medium has the form $E=E_0 \cos(\omega t)$ .Putting this in eqn.(6) we get,

$$P = \varepsilon_0 (\chi^{(1)}E_0 \cos(\omega t)+\chi^{(2)}E_0^2 \cos^2(\omega t)+\chi^{(3)}E_0^3\cos^3(\omega t) +) \tag{7}$$

Using the trigonometric relations,

$$\cos^2(\omega t) = [1+ \cos(2\omega t)] / 2 \text{ and } \cos^3(\omega t) = [\cos(3\omega t) +3\cos(\omega t)] /4 \tag{8}$$

the above relation can be written as,

$$P=(1/2) \varepsilon_0\chi^{(2)}E_0^2+ \varepsilon_0[\chi^{(1)} +((3/4)\chi^{(3)}E_0^2] E_0\cos(\omega t) + (1/2)\varepsilon_0 \chi^{(2)} E_0^2\cos^2 (2\omega t)$$
$$+ (1/4)\varepsilon_0\chi^{(3)}E_0^3\cos(3\omega t)+ \tag{9}$$

The first term is a constant term. It gives rise to a d.c.field across the medium, the effect of which is of comparatively little practical importance. The second follows the external polarization and is called first or fundamental harmonic of polarization; the third oscillates at a frequency 2ω and is called the second harmonic of polarization; the fourth is called the third harmonic of polarization and so on. These higher harmonics of the dielectric polarization, results from the optical nonlinearity of the medium, leads to various optical nonlinear phenomena. Some of these effects are discussed below.

Second Order Nonlinearity

In the case of anisotropic materials (e.g.: uniaxial crystals), all the quadratic, cubic and higher order terms will be active. However, generally the cubic and higher order terms are substantially smaller than the second order term and may be ignored. For such materials, we can write

$$P = \varepsilon_0(\chi^{(1)}) E + \chi^{(2)} E^2) \tag{10}$$

and the medium is said to have second order nonlinearity.

In anisotropic materials, the polarizability depends on the direction of the beam propagation, polarization of the electric field and the relative orientation of the optic axis. Since the vectors P and E are not necessarily parallel in them, the coefficients must be treated as tensors. The second order polarization, therefore may be represented by a relation of the type

$$P_i^{(2)} = \varepsilon_0 \Sigma \chi^{(2)}{}_{ijk} E_j E_k \tag{11}$$

where, i,j,k represents the coordinates x,y,z. However most of the coefficients χ_{ijk} are usually zero so that normally only one or two components need be taken into account.

Second harmonic Generation

Second Harmonic Generation (SHG) is an important application of the second order optical nonlinearity. It can be shown that when two beams of frequencies ω_1 and ω_2 interact with a second order nonlinear medium, the resultant polarization contains the sum $(\omega_1 + \omega_2)$ and the difference $(\omega_1-\omega_2)$ frequencies as given through the expression $2\varepsilon_0\chi^{(2)}E_1E_2$ [cos $(\omega_1+\omega_2)t +(\omega_1-\omega_2)t]$. SHG is the sum of frequency generation when $\omega_1=\omega_2$. The second harmonic beam thus produced is emitted in the same direction as the incident beam.

Third Order Nonlinearity

In case of centro-symmetric materials, i.e., the materials that exhibit inversion symmetry (liquids, gases atomic or molecular vapors etc.) the electric polarization can contain only odd powers of electric field amplitudes. Considering the lowest non-vanishing nonlinear polarization, which is a cubic term in the applied field magnitude, hence we can conclude

$$P = \varepsilon_0 \chi^{(1)}E + \varepsilon_0 \chi^{(3)}E^3 \tag{12}$$

The third order susceptibility term $(\varepsilon_0 \chi^{(3)}E^3)$ will be present with varying strengths in essentially all optical materials irrespective of structural symmetry. Using the above equation, the electric displacement D in the medium can be written as

$$D = \varepsilon_0 E + P = \varepsilon_0(1+\chi^{(1)}) E + \varepsilon_0\chi(3)E^3 = \varepsilon_0(1+\chi^{(1)} + \chi^{(3)}E^2) E \tag{13}$$

Now, the dielectric constant is a nonlinearly varying quantity given by $\varepsilon = \varepsilon_1+ \varepsilon_0\varepsilon_2 \ E^2$ where,

$$\varepsilon_1=\varepsilon_0(1+\chi^{(1)}) : \textit{linear dielectric constant} \tag{14}$$

$\varepsilon_2E^2 = \chi^{(3)}E^2$: nonlinear change in dielectric constant produced by the applied field.

Since, the optical index of refraction n is related to the optical frequency value of ε by $n = (\varepsilon/\varepsilon_0)^{1/2}$, we can also view this as a nonlinear dependence of refraction on the applied signal strength given by

$$n=n_0+n_2I \tag{15}$$

where

$n_0 = (\varepsilon_1/\varepsilon_0)^{1/2}$: linear refractive index and

n_2I is the intensity dependent part of the nonlinear refractive index.

The effect depicted by eqn. 15 is known as the optical Kerr effect because of its similarity to the electro-optic Kerr effect. The optical Kerr effect is a self-induced effect in which the phase velocity of the wave depends on the wave's own intensity.

The order of magnitude of n_2 (in units cm^2/W) is 10^{-16} to 10^{-14} in glasses, 10^{-10} to 10^{-8} in organic materials, 10^{-14} to 10^{-7} in doped glasses and 10^{-10} to 10^{-2} in semiconductors. It is sensitive to the operating wavelength and depends on the polarization. Thus for a centro-symmetric nonlinear medium, the absorption coefficient and the refractive index n must be modified to include intensity dependent terms such that $\alpha(I) = \alpha + \beta I$ and $n = n_0 + \gamma I$, where α (cm^{-1}) and n_0 are the linear terms, and I ($W\ cm^{-2}$) is the incident radiation intensity.

The instantaneous two-photon absorption coefficient β ($cm\ W^{-1}$) and the nonlinear refractive index parameter γ ($cm^2\ W^{-1}$) are related to the corresponding nonlinear susceptibility by the equations

$$Im\ \chi^{(3)}\ (esu) = 10^{-7}\ c^2\ n_0^2\beta\ /96\ \pi^2\omega,$$

and

$$Re\ \chi^{(3)}\ (esu) = 10^{-6}\ c\ n_0^2\gamma\ /\ 480\ \pi^2,$$

where c is the speed of light in cm/s and ω is the fundamental frequency in cycles/seconds.

APPLICATION OF NON LINEAR OPTICAL PROPERTIES

The ability to control the intensity of light in a predetermined and predictable manner is a fundamental and important requirement with applications ranging from optical communications to optical computing. Although there are numerous methods that can be used to switch, limit, amplify or modulate the amplitude of an optical signal, all of these may be broadly classified into two groups, viz. dynamic and passive methods. Optical limiters have been utilized in a variety of circumstances where a decreasing transmission with increasing excitation is desirable. These devices can be used for various pulse shaping applications. An optical limiter consisting of a reverse saturable absorber can be used for passive mode locking [27]. The amplitude modulated pulse can be smoothed by an optical limiter [28]. In this application, a long optical pulse with short intensity pikes incident on the limiter will have the spikes preferentially attenuated with respect to average pulse shape.

It is proposed that an optical limiter together with a saturable absorber can be employed for improved pulse compression [27] in which the leading edge of the pulse is preferentially attenuated by the saturable absorber while the trailing edge will be preferentially attenuated by the optical limiter. The latter is activated by the energy absorbed from the leading portion of the pulse resulting in a more temporally coherent symmetric pulse.

Also slow optical limiters finds applications in reducing the background in pulsed IR laser excited photothermal spectroscopy [14, 28]. However the most important application of optical limiters are in eye/sensor protecting coatings especially in optical systems such as direct viewing devices (telescopes, googles, gunsights etc), focal plane arrays, night vision systems etc. All photonic systems including eye have an intensity level above which damage occurs. Using an optical in the system prior to the sensor extends the dynamic range of the sensor and allows the sensor to continue its operation with out which it could be damaged permanently.

A device that uses some form of active feedback accomplishes dynamic control. A photo sensor, which controls an iris that restricts the intensity of light incident on an optical system, is an example for dynamic control. Dynamic devices suffer from a number of disadvantages, such as higher complexity and slower speeds than passive devices. The higher complexity results from the need for multiple components that must communicate with one another. A device designed for dynamic intensity control generally requires a sensor, a processor and an actuation module that require time to operate in a serial manner and to communicate between the modules.

By contrast, passive control is typically accomplished using a nonlinear optical material in which the sensing, processing and actuating functions are inherent. Since the optical control function is a part of the physical characteristics of the material, the speed is not limited by the communication between the individual modules and the device can be potentially very simple and fast. Such devices are crucial for controlling short optical pulses .Two important and distinctly different types of passive devices used to control the amplitude of an optical signal are "all optical switches and optical limiters". An ideal passive optical switch is a nonlinear optical device that is activated at a set intensity or fluence threshold, whereupon the device becomes completely opaque .By contrast, an ideal optical limiter exhibits a linear transmission below a certain threshold intensity, but above this threshold the output intensity is a constant. The response of an optical limiter and an optical switch are shown in figures. These responses are those of ideal devices to ideal optical pulses that are uniform in both space and time. Pulses with realistic temporal and spatial profile will modify these responses. Under realistic conditions the limiter activation threshold is less well defined, and the output fluence will not be perfectly clamped at a constant value. For any realistic switch, the leading edge of a fast optical pulse will pass through the device before activation, yielding a response intermediate between a limiter and an ideal switch.

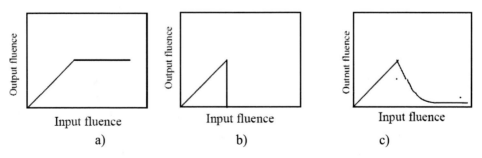

Figure 5. Shows the optical response of (a) an Optical Limiter (b) an ideal Optical Switch (c) A Realistic Optical Switch to the incident fluence.

Mechanisms for Passive Optical Limiting

There are a variety of non-linear optical phenomena that can be used to construct an optical limiter. These include non-linear absorptive process such as two photon (multiphoton), excited state and free carrier absorption; nonlinear refractive processes such as self-focusing and photo refraction and optically induced scattering.

TWO PHOTON ABSORPTION

Two-Photon Absorption (TPA) is an instantaneous non-linearity that involves the absorption of a photon from an electromagnetic. field to promote an electron from its initial state to a virtual intermediate state, followed by the absorption of a second photon that takes the electron to its final real state . Since the intermediate state for such transition is virtual, energy need not to be conserved in the intermediate state but only in the final state. For TPA, the material response is of the order of an optical cycle and is therefore independent of the optical pulse length for a fixed intensity. The device will respond virtually instantaneously to the pulse. On the other hand, because of the low value of the TPA co-efficient β (where $\beta = (3\omega/2 \, \varepsilon_0 \, c^2 \, n_0^{2)})$ Im $[\chi^{(3)}]$ with $\omega =$ the circular frequency of the optical field n_0 –the linear index of refraction and c- the speed of light in vacuum) in usual materials, high intensities are required to realize significant TPA in them . Since the intensity is essentially the energy density divided by the pulse duration, short pulses are required to achieve limiting with TPA for energy densities that may be high enough to damage an optical sensor. Thus, TPA acting alone is not a practical approach to device protection for nano second and longer threat pulses.

REVERSE SATURABLE ABSORPTION

Reverse Saturable Absorption (RSA) or Excited State Absorption generally arises in a system when an excited state absorbs stronger than the ground state. The process can be understood by considering a system that is modeled using three vibronically broadened electronic energy levels. Let the cross section for absorption from the ground state 1 is σ_1, and σ_2 is the cross section for absorption from first excited state 2 to the second excited state 3. The lifetime of the first excited state is σ_2 (seconds). As light is absorbed by the material, the first excited state begins to become populated and contributes to the total absorption cross section. If σ_2 is smaller than σ_1, then the material becomes more transparent and bleaches .Such materials are known as saturable absorber. If σ_2 is larger than σ_1 then the total absorption increases and the material is known as reverse saturable absorber. Reverse saturable absorbers are good optical limiters.

Free Carrier Absorption

Once charge carriers are optically generated in a semiconductor, whether by single photon or two-photon absorption, these electrons (holes) can be promoted to states higher (lower) in the conduction (valence) band by absorbing additional photons. This process is often phonon assisted, although depending on the details of the band structure and the frequency of the optical excitation it may also be direct. The phonon-assisted phenomenon is referred to as free carrier absorption and it is analogous to the excited state absorption in a molecular system. It is clearly an accumulative non-linearity, since it depends upon the build up of carrier population in the band as the incident optical pulse energy is absorbed.

Free carrier absorption always plays some role in the operation of a semiconductor limiter, if the excitation process results in the generation of significant free carrier populations in the band. It certainly contributes to the limiter performance and its inclusion is important in the precise modeling of the response of such devices. Just as in the case of TPA, its importance typically plays in comparisons with non- linear refractive effects, whether single photon or two-photon transition, generates the carriers.

NONLINEAR REFRACTION

Optical limiters based on self-focusing and defocusing form another class of promising devices. The mechanism for these devices may arise from the real part of $\chi^{(3)}$ or from nonlinear refraction associated with carrier generation by either linear or two photon absorption in a semiconductor. Both self focusing and defocusing limiters operate by refracting light away from the sensor as opposed to simply absorbing the incident radiation. Compared to strictly absorbing devices, these limiters can there fore, potentially yield a larger dynamic range before damage to the limiter itself happens.

Here a converging lens is used to focus the incident radiation before it passes through the non- linear medium. The output passes through an aperture before impinging on the detector .At low input levels, nonlinear medium has little effect on the incident beam, and the aperture blocks an insignificant portion of the beam, thus allowing for a low insertion loss for the device. When the nonlinear refraction occurs however the non-uniform beam profile within the medium results in the generation of a spatially non-uniform refractive index. This act as either a negative or positive lens depending on the refractive non-linearity causing the incident beam to either focus or defocus. In a properly designed system this lensing results in a significant amount of energy being blocked by the system aperture thereby protecting the sensor. Since self-focusing can lead to catastrophic damage to the non linear medium itself self defocusing media may have an advantage in practical devices by providing a self protecting mechanism for the limiter itself.

A self-focusing limiter works best if the nonlinear medium is placed approximately at Rayleigh range before the intermediate focus of the device. For a self-defocusing material, the optimum geometry is approximately at one Raleigh range after the focus. This geometric dependence can be exploited to determine not only the sign of the nonlinear refraction in a given medium but the magnitude as well. This is the principle behind the Z- Scan technique [15-20].

Induced Scattering

Scattering is a mechanism that has been extensively studied. Scattering is caused by light interacting with small centers that can be physical particles or simply the interfaces between the groups of non-excited and excited molecules. This scattering can be highly directional or fairly uniform depending on the size of the scattering centers. It is obvious that if an optical signal induces scattering centers in a given medium, the transmission of the medium measured in a given solid angle will decrease. Hence, optical scattering can be used in optical limiters for sensor protection. Induced scattering limiters usually rely on liquid media, because the process in such media is often reversible. That is, if chemical or structural decomposition has not occurred, the excited liquid can readily return to equilibrium. Even when the decomposition does occur, the illuminated volume can be refreshed by either diffusive processes or by circulation. However, when the scattering centers are generated in solids, they are usually due to irreversible decomposition processes that can lead to degradation in the linear operation of the device.

When light impinges on a particle (an atom, molecule or cluster) the electric field interacts with the particle causing the electric charges within to oscillate. The oscillation in turn leads to radiation. This scattered radiation is symmetric with respect to forward and backward scattering. Scattering is of two types: Rayleigh scattering that can be applied to particles much smaller than the wavelength of light or where the particle is nonabsorbing (refractive index real). For particles where size is either comparable or larger than the wavelength of light, then Mie scattering occurs. The essential point is that as the size of the scattering particle increases, a larger percentage of the scattered radiation is forward scattered. Hence, limiting based on Mie scattering will be less effective than Rayleigh scattering [15-20].

Experimental Techniques

Synthesis

Precursor fine particles of magnetite were synthesized by cold co-precipitation from the aqueous solutions of $FeSO_4$ $7H_2O$ and $FeCl_3$ taken in the molar ratio 1 M and 2 M respectively.

Characterization

X-ray diffraction patterns of the dried precursor samples and that of the ferrofluid were recorded in an X-ray diffractometer (Rigaku D max-C) using Cu Kα radiation (λ=1.5406 Å), and planes were identified using the JCPDS tables [29]. The X ray diffraction spectrum of the magnetite ferrofluids is depicted in Figure 6. The average particle sizes of these powder samples were estimated from the Debye Scherrer's formula

$$D = \frac{0.9\lambda}{\beta \cos\theta} \tag{16}$$

where λ is wave length of X-ray used, β is the FWHM of the XRD peak with the highest intensity, and D is the particle diameter.

Non Linear Optical Studies

Non linear optical measurements were carried out using laser pulses of 7 ns as well as 100 fs durations (FWHM). Nanosecond pulses at 532 nm were obtained from a frequency-doubled Nd:YAG laser, while femtosecond pulses at 800 and 400 nm were obtained from a Ti:Sapphire CPA laser (TSA-10, Spectra Physics) equipped with a harmonic generator. Samples were taken in a 1 mm cuvette. The laser pulses were plane polarized with a Gaussian spatial profile. The intensity dependent light transmission through the sample was measured using an automated z-scan set-up. In the z-scan technique, the laser beam is focused using a lens, and the sample is moved along the beam axis (z-axis) from one side of the focus to the other, through the focal point. At each position z, the sample sees a different laser fluence, and the fluence will be a maximum at the focal point (where z is taken as zero). An energy meter placed after the sample measures the position dependent (ie, fluence-dependent) transmission through the sample [30].

In our experiment we added another detector, which is a photo multiplier tube (PMT), to record scattered radiation from the sample. The PMT is kept at a radial distance of 5 cm from the beam axis.

Both the PMT and the cuvette are mounted on the same translation stage, so that they remain at the same relative distance throughout the scan.

For all the three wavelengths used, the samples show a minimum transmission around the beam focus, showing that the nonlinearity is of the optical limiting type. In general, optical limiting in media can arise from different types of processes like nonlinear refraction, two-photon absorption, excited state absorption and induced light scattering. Two-photon absorption is an intensity dependent process, in which two photons of the exciting radiation are absorbed simultaneously. On the other hand, excited state absorption is a sequential process in which an absorbed photon initially raises the molecule to an excited state of finite lifetime, from where it absorbs one more photon to reach a terminal level. In the case of induced light scattering, in transparent media it is usually caused by the microplasma formed due to laser-induced dielectric breakdown. In absorbing colloidal suspensions it can be of thermal origin as well. Here the particles absorb laser light and heat the liquid so that the local temperature exceeds the ambient temperature, resulting in transient refractive index changes. If the temperature exceeds the liquid's boiling point, microbubbles also will be generated.

As a result of these, light will get scattered at high input fluences. Such induced scattering contributes to optical limiting in a number of materials, including carbon black [31] and carbon nanotube [32] suspensions, and metal nanoparticles [33, 34]. The PMT kept off-axis in our z-scan experiment monitors the normal as well as induced scattering from the sample.

RESULTS AND DISCUSSIONS

From the measured line broadening the particle size is calculated to be around 90Å, and the planes are identified. The lattice parameter 'a' has been evaluated assuming cubic symmetry, and its value is found to be 8.314 Å.

The optical absorption spectrum of the magnetite-based ferrofluid is shown in Figure 7. The absorbance gradually decreases through the visible, towards the IR region. A weak absorption band can be seen around 490 nm. In order to calculate the band gap (assuming direct band gap), it may be noted that for a semiconductor, the absorption coefficient near the band edge is given by

$$\alpha = \frac{A(h\nu - E_g)^{1/2}}{h\nu} \qquad (17)$$

where A is a constant and E_g is the energy band gap. When $\alpha h\nu = 0$, $E_g = h\nu$, and therefore, using an extrapolation the energy bandgap is determined to be 3.1 eV.

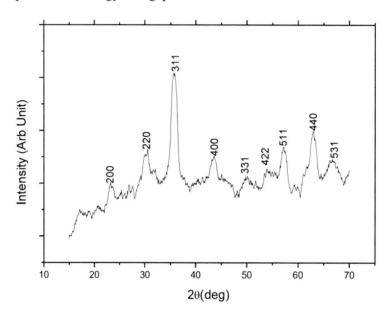

Figure 6. X Ray Diffraction Spectrum of magnetite ferrofluids.

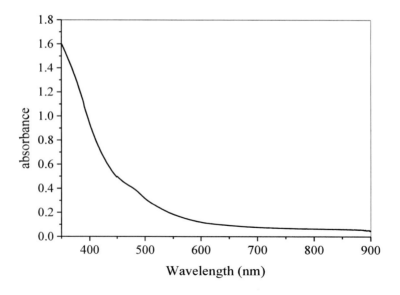

Figure 7. Linear absorption spectrum of the ferrofluid sample FF1.

The causes of limiting being so diverse, we have studied the samples at two different concentrations, in addition to the pure base fluid. Samples having linear transmissions of 0.5 and 0.75 at the respective wavelengths are employed for all measurements. When excited with low energy nanosecond pulses, the base fluid kerosene shows very little limiting and the ferrofluid suspensions show a moderate limiting behaviour.

Pure kerosene exhibits weak two-photon absorption, and the two-photon absorption coefficient β is calculated to be equal to 5 x 10^{-12} m/W from numerical fits to the experimental data. However, data from the samples seem to fit to a three-photon absorption process,

$$T = \frac{(1-R)^2 \exp(-\alpha L)}{\sqrt{\pi} p_0} \int_{-\infty}^{+\infty} \ln\left[\sqrt{1 + p_0^2 \exp(-2t^2)} + p_0 \exp(-t^2)\right] dt \tag{18}$$

where $p_0 = [2\gamma(1-R)^2 I_0^2 L_{eff}]^{1/2}$. Here T is the transmittance, R is the Fresnel reflection coefficient at the sample-air interface, α is the absorption coefficient, L is the sample length, and I_0 is the peak on-axis intensity incident on the sample.

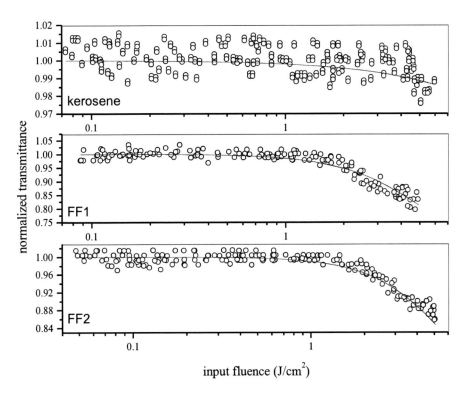

Figure 8. Fluence dependent transmission of the base fluid and the samples at 7ns laser pulse excitation. Excitation wavelength is 532 nm, and the maximum laser fluenceused is 6 J/cm^2. The linear transmissions of FF1 and FF2 are equal to 0.5. Kerosene is transparent at this wavelength. Circles denote experimental data, and the solid curves are theoretical fits (see text).

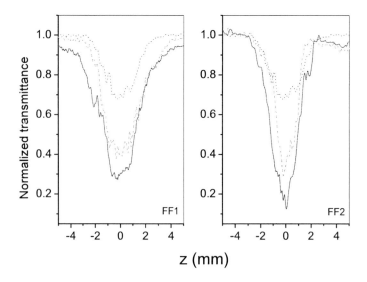

Figure 9. Concentration dependence of the Z-scans obtained for the ferrofluid samples at 7ns laser pulse excitation. Excitation wavelength is 532 nm. Laser fluence is approximately 20 J/cm² at z=0. Sample linear transmissions are, Solid line – 0.5, and Dashed line – 0.75. Dotted line – pure kerosene.

L_{eff} is given by $[1-\exp(-2\alpha L)]/2\alpha]$. The three-photon absorption coefficients γ are found to be 7.5 x 10^{-23}m³/W² and 5 x 10^{-23}m³/W² respectively, for FF1 and FF2. The above values of β and γ are applicable in the limit of moderate laser excitation, at the specified concentrations. As seen from Figure 9, at higher fluences a stronger, concentration dependent limiting effect is observed.

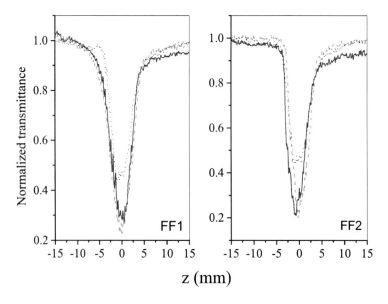

Figure 10. Z scan curves obtained for the ferrofluid samples at 100 fs, 800 nm laser pulse excitation. Sample linear transmissions are, Solid line – 0.5, and Dashed line – 0.75. Dotted line – pure kerosene.

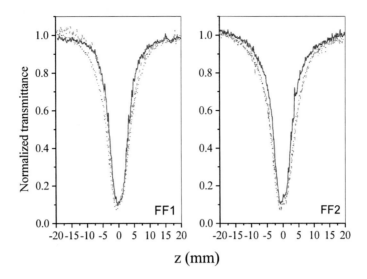

Figure 11. Z scan curves obtained for the ferrofluid samples at 100 fs, 400 nm laser pulse excitation. Sample linear transmissions are, Solid line – 0.5, Dashed line – 0.75. Dotted line – pure kerosene.

Figure 10 shows the Z scan curves obtained with femtosecond pulses, for the excitation wavelength of 800 nm. Limiting in the base fluid is substantially increased, though it is less than that exhibited by the samples. However, when the excitation wavelength is changed to 400 nm, there is practically no difference between the limiting efficiency of the pure base fluid and the ferrofluid suspensions (Figure 11).

Figure 12 gives the signals recorded by the PMT as a function of the sample position z. When the sample is away from the beam focus a background scattering can be seen, which is a minimum for pure kerosene and more for the ferrofluid suspensions.

In the nanosecond regime the signal amplitude goes up near the beam focus, indicating the occurrence of induced scattering. In the pure base fluid this may be caused by the microplasma, whereas in the samples it can also be of the thermal origin. The thermally induced transient refractive index change is given by

$$\Delta n_0 = (dn_0/dt)F_0\alpha/2\rho C_v \qquad (19)$$

where dn_0/dt is the thermo-optic coefficient, F_0 is the laser fluence, ρ is the density and C_v is specific heat at constant volume. The induced scattering amplitude is found to be the highest in the base fluid, and lesser in the suspensions, even though optical limiting in the suspensions is substantially higher than that seen in the base fluid (Figure 9). Therefore nanosecond optical limiting in the ferrofluids cannot be fully accounted for by induced scattering. Moreover, numerical fitting to the data (Figure 8) shows that in the samples an effective fifth order nonlinearity is manifested even at relatively lower pump fluences. Since there are no compelling reasons for a higher order process like three-photon absorption to be dominant in these systems, what is more likely is that there are sequential excited state absorption events, in which free carriers take an active part. The collective effect of such events will appear in the form of higher order processes. For example, a $\chi^{(5)}$ nonlinearity of the $\chi^{(3)}:\chi^{(1)}$ type has been observed in semiconductors [35, 36]. For metal nanoparticles, photoexcitation of free carriers has been discussed previously by Kamat et.al. [18], and the role of free carriers in enhancing their nonlinear absorption also has been discussed [37]. Similar behaviour in molecular metal clusters is also reported [38].

In the femtosecond excitation case the PMT signal behaves differently, as the ultrashort pulse is too fast to register any thermal scattering events.

Some induced scattering is observed in the base fluid, but the suspensions do not show any nonlinear increase in the scattering level. In fact at the highest intensities, rather a decrease in the signal level is observed. Details are given elsewhere [39]. The reason for this decrease is not well understood now, but it clearly shows that induced scattering does not contribute to femtosecond optical liming under the present experimental conditions. Considering the very short timescale and the high intensities involved, instantaneous nonlinearities like two-photon and multiphoton absorption should be responsible for the limiting. As seen from Figure 11 limiting is the best in the pure base fluid, and the addition of the ferrofluid does not improve the inherent optical limiting of kerosene in the femtosecond regime.

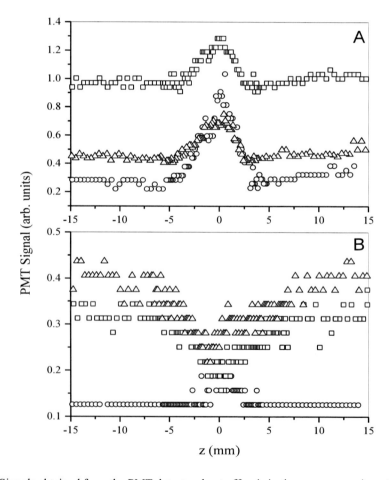

Figure 12. Signals obtained from the PMT detector, kept off-axis in the z-scan experiment. Plot A is for nanosecond excitation (532 nm) and plot B for femtosecond excitation (400 nm). Circles denote pure base fluid (kerosene), squares are for FF1, and triangles for FF2. *Both FF1 and FF2 have a linear transmission of 0.5.*

CONCLUSION

Magnetite based ferrofluids synthesized by co-precipitation technique with oleic acid as the surfactant and kerosene as the carrier, are characterized by optical absorption and X ray diffraction measurements. The absorptive nonlinearities have been studied, and the wavelength, fluence, concentration and pulse width dependence of the nonlinear transmission have been investigated. Results indicate that induced scattering and excited state absorption contribute to the observed optical limiting in the nanosecond excitation regime. On the other hand, in the femtosecond regime, pure electronic effects are responsible for the limiting. Due to their good thermal stability, resistance against agglomeration and long shelf life, ferrofluids will be potential candidates for optical limiting applications.

REFERENCES

[1]   E. Ronald Rosenweig, Ferro hydrodynamics, Cambridge University Press 1985.

[2]   B.M. Berkovsky, V.S. Medvedev, M.S. Krakov, Magnetic fluids; *Engineering applications,* Oxford university press. 1993.

[3]   R.K Bhat, *Indian J. Eng. Mater. Scie.* 5, (Dec 1998), 477.

[4]   J. Smit and H.P.G Wijn, Ferrites (Philips Technical Library) 1959.

[5]   J.Deperiot, G. J Da Silva C.R Alves *Brazilian J. of Phys.* 31, (2001), 390.

[6]   E. Hasmonay, J Depeyrot, *J. Appl.Phys.* 88, No 11 (Dec. 2000), 6628.

[7]   G. M Sutharia and R.V Upadhyaya *Indian J. Eng. Mater.Scie.* 5, (1998), 347.

[8]   K.T. Wu, P.C. Kuo, Y.D. Yao and E.H Tsai, *IEEE trans. Magn,* 37 (2001). 2651.

[9]   HelenW.Davies and Patrick Llewellian, *J. Phys. D, Appl. Phys,* 12, 1979, 1357.

[10] S Couris, E. Koudoumas, A A Ruth and S. Leach, *J.Phys.B, At.Mol.Opt.Phys.* 28 (1995) 4537.

[11] L. W. Tutt and A. Kost *Nature* 356 (1992), 225.

[12] Kost, L. W. Tutt, M. B. Klein, T. K. Dougherty, and W. Elias, *Opt. Lett,* 18 (1993) 334.

[13] D. G. Mc Lean, R. L. Sutherland, M.C. Brant, D. M. Brandelik, P. A. Fleitz and T. Pottenger, *Opt Lett,* 18 (1993) 858.

[14] F. Lin, J. Zhao, T. Luo, M. Jiang, Z. Wu, Y. Xie, Q. Qian , H. Zeng *J. Appl. Phys,* 74 (1993) 2140.

[15] B. L. Justus, Z. H. Kafafi, A. Huston L, *Opt. Lett,* 18 (1993) 1603.

[16] R.Philip, G.Ravindra Kumar, N.Sandhyarani, T.Pradeep, *Phys.Rev.B.* 62 (2000) 13160.

[17] S.Link, C.Burda, Z.L.Wang, M.A.El-Sayed, *J.Chem.Phys.* 111 (1999) 1255.

[18] P.V.Kamat, M.Flumiani, G.V.Hartland, *J.Phys.Chem.B.* 102 (1998) 3123.

[19] C.Zhan, D.Li, D.Zhang, W.Xu, Y.Nie and D.Zhu, *Opt.Mat.* 26 (2004) 11.

[20] K.S.Bindra, S.M.Oak and K.C.Rustagi, *Opt.Commun.* 124 (1996) 452.

[21] H.P.Li, C.H.Kam, Y.L.Lam, and W.Ji, *Opt.Commun.* 190 (2001) 351.

[22] A.A.Said, M.Sheik Bahae, D.J.hagan, T.H.Wei, J.Wang, J.Young, and E.W.Van Stryland, *J.Opt.Soc.Am.B.* 9 (1992) 405.

[23] Y.B Band, D.J Harter and R. Bavli, *Chem. Phys. Lett,* 126, (1986) 280.

[24] S.E Bialkowski, *Opt. Lett,* 14, (1989) 1020.

[25] H.W. Davies and P.Llewellian, *J. Phys. D, Appl. Phys,* 12, 1979, 1357.

[26] S. Couris, E. Koudoumas, A. A. Ruth, S. Leach, *J.Phys.B, At.Mol.Opt.Phys.* 28 (1995) 4537.

[27] D.J Harter and Y.B Band, Springer Series, *Chem. Phys*, 38, (1984) 102.

[28] D.J Harter, M.L Shand and Y.B Band, *J. Appl. Phys*, 56, (1984) 865.

[29] JCPDS-ICDD C 1990 10-319.

[30] M. Sheik Bahae, A. A Said and Van Stryland, *Opt. Lett.* 14, (1989) 955.

[31] L. W. Tutt, T. F. Boggess, *Prog.Quant.Electron.* 17 (1993) 299.

[32] S. R. Mishra, H. S. Rawat, S. C. Mehendale, K. C. Rustagi, A. K. Sood, R. Bandopadhyay, A. Govindaraj, C. N. R. Rao, *Chem.Phys.Lett.* 317 (2000) 510.

[33] L.Francois, M.Mostafavi, J.Belloni, J.F.Delouis, J.Delaire, P.Feneyrou, *J. Phys.Chem.B.* 104 (2000) 6133.

[34] Anija M., Jinto Thomas, Navinder Singh, Sreekumaran Nair, Pradeep T. and Reji Philip, *Chem. Phys. Lett.* 380, 223 (2003).

[35] *Handbook of Nonlinear Optics*, by R.L.Sutherland, Marcel Dekker (NY) 1996.

[36] M. Sheik Bahae, A. A Said and Van Stryland, *IEEE J.Quant.Electron.*

[37] C.Zhan, D.Li, D.Zhang, W.Xu, Y.Nie and D.Zhu, *Optical Materials* 26 (2004) 11.

[38] B.Karthikeyan, Jinto Thomas, Reji Philip *Chem. Phys. Lett.* 414, 346 (2005).

[39] Swapna S. Nair, Jinto Thomas, C. S. Suchand Sandeep, M. R. Anantharaman, and Reji Philip, *Appl. Phys. Lett.* **92**, (2008) 171908.

In: Ferrofluids
Editors: Franco F. Orsucci and Nicoletta Sala

ISBN: 978-1-62808-410-8
© 2013 Nova Science Publishers, Inc.

Chapter 6

SUPERPARAMAGNETIC IRON OXIDE NANOPARTICLES BASED AQUEOUS FERROFLUIDS FOR BIOMEDICAL APPLICATIONS

Mary A. P. Reena and M. R. Anantharaman
Department of Physics, Cochin University of Science and Technology,
Cochin, India

ABSTRACT

This chapter deals with the synthesis of highly stable water based iron oxide fluid with narrow particle size distribution at neutral pH, and the evaluation of magnetic properties for hyperthermia application. The power loss spectrum of these nanoparticles in an external alternating magnetic field is simulated to investigate the possibility of applying in AC magnetic heating. The nanoparticles are modified with silica coating that reduces the particle interaction. The cell viability test conducted with these fluids on He La cells was promising and it showed enhanced biocompatibility on silica modification. An introduction is provided for better understanding of the readers.

Bio-compatible magnetic fluids having high saturation magnetization find immense applications in various biomedical fields. Aqueous ferrofluids of superparamagnetic iron oxide nanoparticles with narrow size distribution and high shelf life is realized by a controlled chemical co-precipitation process. Particle sizes are evaluated by employing transmission electron microscopy. Room temperature and low temperature magnetic measurements were carried out with a superconducting quantum interference device. The fluid exhibits good magnetic response even at very high dilution (6.28mg/cc). Based on the structural and magnetic measurements, the power loss for the magnetic nanoparticles under study is evaluated over a wide range of radio frequencies. The iron oxide nanoparticles were embedded in silica synthesized through Stober process, and suspended in polyethylene glycol. The nanoparticles exhibited non interacting nature and enhanced bio-compatibility.

INTRODUCTION

Colloidal suspensions of ultrafine magnetic particles (ferrofluids) have widespread applications in fields such as biology and biomedicine [1-4]. Ferrofluids are synthesized by dispersing nanosized magnetic particles in carrier liquids with suitable surfactants and proper stabilization techniques. The biocompatibility and the ease with which it can be dispersed in water qualify iron oxide based ferrofluid a competent candidate for membrane separation, intraocular retinal repair, early diagnosing, imaging and magnetic hyperthermia for cancer therapy [5], enzyme immobilization of cell targeting, targeted drug delivery [6], in smart ferrogel preparation [7] for controlled delivery of drugs and as magnetic resonance imaging (MRI) contrast enhancing agents [8]. Magnetic nanoparticles–drug conjugate attached to an antibody or hormone can be magnetically guided to the infected tissue and could specifically bind to it while the external magnetic field is on. As the field is removed, the drug will disperse in the whole body, but the amount of drug is small, there will be very little side effects than when the drug is taken to the whole body initially. This provides a platform for optimum dosage of drug. Surface modified magnetic nanoparticles are applied in cell separation such as clean up of bone marrow [9]. This includes fixing of monoclonal antibodies (that has affinity towards the tumor cells), coated magnetic particles to the cells. The tumor cells specifically attach to the nanoparticles and are magnetically separated in a field gradient. The specificity of the nanoparticles could be provided by surface modification with suitable functional molecules.

MRI is one of the powerful techniques in medical imaging, which measures the spin lattice relaxation time (T_2), the time that a proton in water takes to achieve thermodynamic equilibrium with the surroundings. T_2 varies for different tissues and when a magnetic particle is loaded to a tissue, the magneto static field generated by the magnetic nanoparticles can alter the relaxation of proton enhancing the contrast. Iron oxide nanoparticles coated with dextran are biocompatible and is used as contrast enhancing agents [10]. High magnetic moment is required for a contrast agent. The surface area, size and shape of the nanoparticles decide the physical and chemical properties of these particles to a great extent, which in turn decide the performance in various applications [11]. The Brownian relaxation depends on the hydrodynamic volume, which varies with the molecules attached [12] to the magnetic particle. This property can be used for biomolecular sensing. Any shift in relaxation time reflects the properties of the binding and the dynamics of binding [10]. The particle size, and its distribution along with the magnetic and flow properties of the fluid influence the application parameters in biomedicine. The spherical shape and mono dispersibility of SPIONs are often a prerequisite for applications in living tissues [13], since any small deviation in hydrodynamic volume can change the relaxation characteristics, thus leading to misinterpretations. Moreover the particle, the surface coating and the carrier need to be bio-compatible. Thus the optimization of the synthesis of nanoparticles and their conjugation with organic molecules onto the surface becomes very much essential. Aqueous based ferrofluids are the best candidates for practical approach in biomedicine.

Biosensing [14] is another area where the magnetic nanoparticles (MNPs) can be employed. The MNPs are applied in diagnostic techniques where the pathogens or cells in biological samples are sensitively measured. This includes early detection and diagnosis of tumors, monitoring malignancy and sensing the efficacy of treatment. Diagnostic magnetic resonance (DMR) is a technique in which the MNPs act as proximity sensors, which modulates the spin-spin relaxation of the water molecules surrounding the nanoparticles due to the strong field produced by them.

Hyperthermia refers to the degeneration of living tissues by heating the cells. This technique is applied to destroy the malignant cells in cancer treatment. Magnetic hyperthermia is superior to other techniques of hyperthermia for cancer treatment because of reduced side effects such as damage to healthy tissues [15]. Magnetic hyperthermia or magneto-thermo cytolysis refers to heating of cells attached to magnetic particle by an external AC magnetic field. The increase in temperature is caused due to hysteresis loss/ Neel relaxation. In the case of superparamagnetic particles, the loss is caused by the relaxation processes [16]. The heat dissipated when subjected to an alternating magnetic field depends on the fluid properties such as viscosity, the ratio of relaxation frequency to the applied frequency, distribution of the magnetic component, domain magnetization and density and the specific heat capacity of the magnetic constituent [17]. The heat produced in external magnetic field raises the temperature of the tissue and this causes the rupture of cells.

The stability of the fluid against sedimentation is decided collectively by competing interactions [1, 18-20] such as van der Waals interactions, dipolar interactions, viscous force of the carrier liquid, and the electrostatic and steric repulsion of the surfactant. Surfacted ferrofluid have a long chain of organic molecule such as oleic acid and oleilamine, around the surface and mainly, steric repulsion provides stabilization. In ionic fluids, the electrostatic repulsion provides stabilization. Hence the pH of such fluids may vary considerably (from 3 to 9) from acidic to basic depending on the surface treatment of the nanoparticles after precipitation. Biomedical applications require the fluid to be in neutral pH. Citric acid, a biocompatible surfactant presents both electrostatic and steric effects and could easily get conjugated to iron oxide particles. Iron oxide is most recommended because of its higher magnetization values, lesser toxicity [21] and the ease of metabolism by the liver.

The magnetic property of the nanoparticles is decided by the intrinsic magnetic parameters such as magnetization, coercivity and the inter-particle interaction. The interaction of particles modifies the relaxation mechanism [22] and coercivity and this affects the application in living cells. The surface modification of magnetic nanoparticles with inorganic materials like silica can reduce the inter-particle interaction. Moreover the silanol groups present on the surface of silica can bind to different functional groups easily and is highly recommended for bio-functionalisation. The functionality of these chemical species is such that they have affinity to cancerous cells than normal healthy ones. With the surface modification by silica the total size of the particles increases and the stability of suspension demand a different carrier liquid other than water, or a different pH. Polyethylene glycol (PEG) is biocompatible [23] with higher viscosity and can suspend silica modified SPIONs. The iron oxide particles larger than 50 nm are utilized in *in vitro* magnetic separation purposes due to their ferromagnetic behavior, while for *in vivo* biomedical applications iron oxide nanoparticles less than 50 nm are used. Hence, size control over the silica shell is important in the case of the SPION based silica composites for different biomedical applications [22]. The bare nanoparticles due to their smaller size may undergo phagocytosis by the nearby tissues [24]. The silica modification can prevent this as well. The biocompatibility, ease of functionalisation and stability makes the silica coated nanoparticles and their ferrofluids superior for biomedical applications. This core shell structures of iron oxide in silica assemblies are interesting in understanding the fundamental magnetic properties of nanomaterials.

Synthesis

Synthesis of Aqueous Ferrofluids

Monodispersed iron oxide particles of average size 9.5 nm were synthesized through controlled chemical co-precipitation method. For this, analytical grade anhydrous ferric chloride (FeCl$_3$) and ferrous sulphate hepthahydrate (FeSO$_4$.7H$_2$O from Merk) in the molar ratio of 2:1, each in 500ml of distilled water were taken as the starting solution. 12% of aqueous ammonia was added to the solution while stirring at room temperature to supersaturate for the precipitation of the oxide. The rate of reaction was controlled by allowing one drop of ammonia per second to react with this solution until a pH of 10, to get a thick dark precipitate. 5 g of citric acid (COOH-CH$_2$-C(COOH)(OH)-CH$_2$-COOH) crystals dissolved in 10 ml water was added to this wet precipitate and allowed for further reaction at an elevated temperature of 80°C while stirring for another 90 minutes. This sample was then washed with distilled water several times for the removal of water soluble byproducts. This is then suspended in distilled water by ultrasound treatment. The obtained fluid was kept for gravity settling of any bare nanoparticles and was then centrifuged at a rotation speed of 3500 rpm to remove any particles that may sediment. The supernatant fluid is extracted for further analysis. The concentration of the magnetic particles is estimated to be 6.28 mg/cc.

Synthesis of Silica-Coated PEG-Based Ferrofluid

A modified sol-gel synthesis that involves hydrolysis and condensation of TEOS was employed to obtain silica-coated iron oxide ferrofluids. In order to achieve this, 2 g of the above synthesized CA-coated iron oxide ferrofluid was diluted in 10 ml of distilled water. In addition, 5 ml of NH$_4$OH and 30 ml of ethyl alcohol were added to this solution. This dispersion was homogenized by ultrasonic vibration for 10 minutes. Under continuous and uniform stirring, 1 ml of TEOS was slowly added to this mixture and the temperature was maintained at 80°C. The resulting precipitate was magnetically decanted. A schematic of the synthesis is presented in (Figure 1).

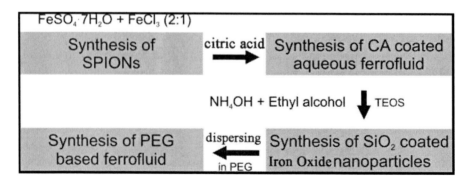

Figure 1. Schematic of silica modified SPIONs suspension in PEG.

One part of the precipitate was dispersed in water followed by extensive ultrasonication while the other part was dispersed in viscous and biocompatible fluid, PEG (viscosity=1.6×10^{-2} Pa.s; density=1.124g/cc. PEG employed for the present study is PEG 400 (Merck Chemicals, Analytical Grade) with the structure HO-CH$_2$-(CH$_2$-O-CH$_2$-)n-CH$_2$-OH and it has n=9. It was found that the water dispersed SiO$_2$-coated iron oxide nanoparticles (particle size is about 9 nm; SiO$_2$ shell is about 280 nm) settle down quickly (Figure 2a), while dispersed in PEG, SiO$_2$-coated iron oxides were found to be stable even after 6 months of ageing or under a small external magnetic field (Figure 2b).

One of the possible mechanisms for the formation of PEG-based stable ferrofluids is schematized in Figure 2c. Silica shells were formed by the Stober's process [25] within the iron oxide ferrofluid followed by the addition of TEOS in the presence of ethanol and concentrated ammonium hydroxide. The hydrolysis, condensation and polymerization take place leading to the formation of silica nanospheres. During the process of silica nanospheres formation, SPIONs are entrapped within silica shell and this is equivalent to a core-shell structure with SPIONs as the core and silica as the shell. More than one iron oxide nanoparticle can be entrapped within a silica shell. PEG can easily bind with silica through the -OH groups giving rise to stable dispersions. Viscous PEG provides enhanced force of buoyancy for silica-coated magnetic particles than that in water (viscosity of water 8.9×10^{-4} NS/m^2). The pH of the PEG based ferrofluid was found to be ~7.

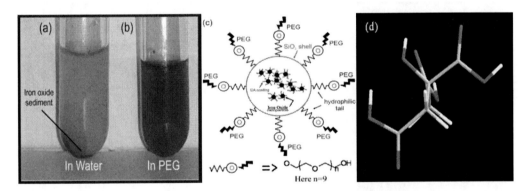

Figure 2. Photograph of the silica modified nanoparticles suspended in (a) Water (b) PEG (c) schematic of silica encapsulated nanoparticles (d) Structure of citric acid.

The structural characterization is carried out using X-ray diffraction (XRD) technique (Rigaku D Max) at Cu Kα. The particles are analysed with transmission electron microscopy (TEM). The samples for the above mentioned experiments are prepared by evaporating the moisture content from the fluid. Room temperature and low temperature magnetic measurements are performed in a Superconducting Quantum Interference Device (SQUID) magnetometer (MPMS Quantum Design). The simulation of the power loss spectrum of the sample is performed for applied field strength of 500 Oe in the range 100 kHz to 1000 kHz. Scanning electron microscopy (SEM, JEOL, JSM-6390LV), energy dispersive spectroscopy (EDS) Fourier transform infrared spectroscopy (FT-IR, Thermo Nicolet Avatar 370 DTGS).

Cytotoxicity Assay

Cytotoxicity of citric acid coated aqueous ferrofluid and silica-coated PEG-based ferrofluid are evaluated by 3-(4,5-dimethylthiazole-2-yl)-2,5-diphenyltetrazolium chloride

(MTT) assay[39] Here, ~1x10^6 HeLa cells[*] are inoculated into 96 well tissue culture plates containing DMEM supplemented with 10% FBS and incubated for 48h at 37^{0}C. The cells are copiously washed with PBS, and the medium is exchanged with DMEM containing different concentrations of ferrofluids. The cells are incubated for 24 hr at 37°C and subjected for MTT assay following standard protocol. Briefly, ferrofluid treated cells are supplemented with 50 µl of MTT solution (5 mg mL^{-1}) prepared in PBS and kept for incubation under dark at 37 ^{O}C for 5hr. Subsequently, the viability of the cells is measured as a function of reduction of yellow MTT to insoluble purple formazan by mitochondrial dehydrogenase enzyme of healthy cells. Formazan crystals were dissolved in Dimethyl Sulphoxide (DMSO, HiMedia) and the absorbance was recorded at 570 nm using a microplate reader (InfiniteM-200 Tecan, Austria). Each experiment is repeated six times, and mean and standard error were calculated.

RESULTS AND DISCUSSION

The fluids exhibited good shelf life and stability against sedimentation under gravitational and magnetic fields. When an external magnetic field is applied normal to the free surface, fluid of citric acid coated SPIONs exhibited good spiking.

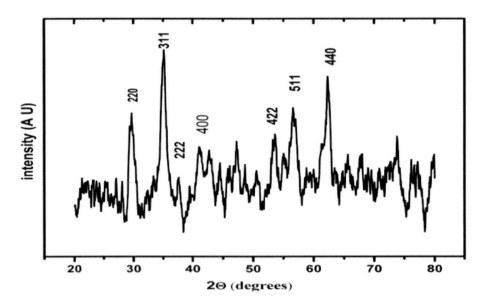

Figure 3. XRD pattern of the SPIONs.

Structural and Morphological Analysis

The X-ray diffraction pattern (Figure 3) shows that the iron oxide particles have crystallized in the inverse spinel structure with a lattice constant of 8.41Å. All the major crystallographic planes corresponding to inverse spinel are identified [ICDD PDF No.750449]. It is hard to differentiate between maghemite (γ Fe$_2$O$_3$) and magnetite (Fe$_3$O$_4$) using X-ray diffraction analysis alone, since both represent an inverse spinel structure and the (hkl) planes are similar. However from XRD analysis it is seen that the compound contains no traces of nonmagnetic haematite (α-Fe$_2$O$_3$). There is significant broadening of peaks due to the size reduction of particles. The particle size is calculated from the line broadening applying Scherrer's formula, and is found to be 9.5 nm. The TEM images (Figure 4a, b) show that the particles are nearly spherical. Statistical analysis (Figure 4c) of the images revealed a normal distribution of particles with a mean size of 10 nm and a width of 3 nm. This is in fair agreement with the crystallite size obtained from X- ray diffraction measurements. The energy dispersive spectrum shows only signatures of iron, showing the purity of SPIONs.

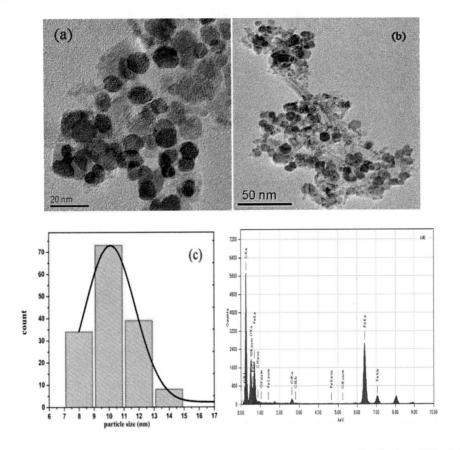

Figure 4. (a,b): Transmission electron micrographs (TEM), 2(c) particle size distribution, 2(d) EDS of FFCA.

The narrow size distribution is an advantage while considering the magnetic hyperthermia applications or for targeted drug delivery. The hydrodynamic length of a single citric acid molecule is calculated to be nearly 0.7 nm. This is the thickness of the surfactant monolayer.

So there is a minimum spacing of 1.4 nm between the iron oxide nanoparticles. This could result in dipolar interactions among the particles. The dipolar interactions can be brought down if a surface coating on the magnetic particles is possible. This is realized by silica modification of magnetic nanoparticles.

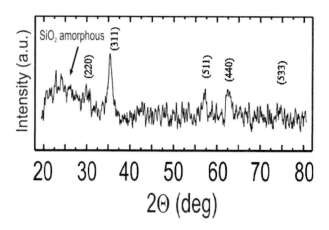

Figure 5. X-ray diffraction pattern of silica coated iron oxide.

Figure 5 shows the XRD pattern of silica modified SPIONs. A broad peak is seen at around 20° corresponding to amorphous silica. Particle size of iron oxide calculated from the diffraction peak is about 9.3 nm, which shows that the particles have not grown while silica is coated. The nature of bonding in the PEG-based ferrofluid was further investigated using FT-IR (Figure 6).The spectrum of SiO_2-modified iron oxide nanoparticles dispersed in PEG has characteristic bands of both silica and iron oxide [26]. The spectrum shows the characteristic peaks of Fe-O (673 cm^{-1}), C-O (from PEG), Si-O-Si and Si-OR (a broad peak centered at 1078 cm^{-1}) bonds. The peaks at 885 cm^{-1} and 944 cm^{-1} can be assigned to symmetric and asymmetric stretching vibrations of the terminal groups of Si-OH and Si-O-R. The frequency mode which is at 1078 cm^{-1} is associated with the back and forth motion of the oxygen atom along a line parallel to the Si-Si axis. This motion results in the opposite distortion of two neighboring Si-O bonds. A weak band is observed at 830 cm^{-1} due to symmetric stretching of the oxygen atom along a line bisecting the Si-O-Si angle. The absence of a band at 857 cm^{-1} in the spectrum suggests that there is no Si-O-Fe bond in the composite. The latter indicates that iron oxide is in an isolated state. Moreover, the presence of silanol bonds establishes the hydrophilic nature of silica shell and hence this also confirms the formation of the stable ferrofluid with PEG as carrier fluid. The PEG based ferrofluid remained stable even after an ageing for six months.

The SEM and TEM images of the silica-coated iron oxide nanoparticles are shown in Figure 7. The SEM image confirms the formation of white spherical particles having an average size of 280 nm (Figure 7a). The enhanced size is attributed to the silica coating.

Figure 7b depicts the bright field TEM image of silica-modified iron oxide nanoparticles. The contrast difference indicates that iron oxide and silica forms core-shell structures. Also it is to be noted that a single silica shell contains more than one iron oxide nanoparticle.

The size of the iron oxide nanoparticles found in TEM is ~10 nm and correlates well with that calculated from XRD. The elemental analysis using EDS was carried out revealing the presence of Si, Fe and O. The molar mass ratio between iron oxide and Silica, calculated from the relative atomic percentage of Fe and Si from EDS, is found to be 1:17. Figure 8a shows the bright field TEM image of silica-modified iron oxide nanoparticles for which elemental mapping has been carried out and is shown in Figure 8b-d.

Superparamagnetic Iron Oxide Nanoparticles Based Aqueous Ferrofluids ... 85

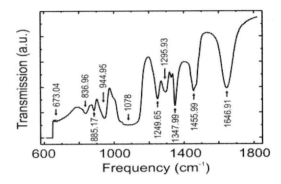

Figure 6. FTIR spectrum on silica modified iron oxide.

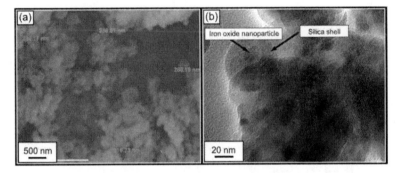

Figure 7. (a) SEM image, (b) TEM image of the silica coated iron oxide.

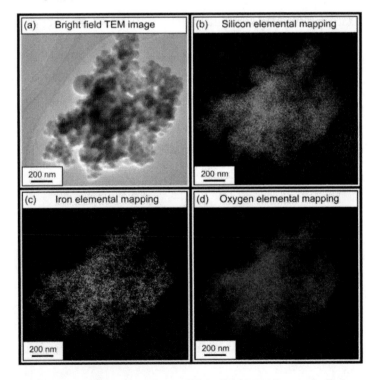

Figure 8. SiO$_2$-coated Iron Oxide nanoparticles (a) bright field TEM image. (b-d) elemental mapping of Si, Fe and O, respectively.

It is to be noted that uniform colour distribution was obtained for Fe, Si and O indicating the uniform distribution of core-shell nanostructure.

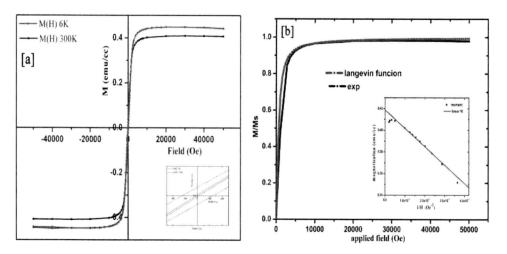

Figure 9. (a) Hysteresis loops of FF CA particles; inset enlarged view at low applied fields (b) Theoretical fitting of normalized moment with Langevin function (Inset) Magnetization - H^{-1} plot for iron oxide nanoparticles.

Magnetic Characterization

For magnetization measurements, 3 micro litres of the fluid was dropped over to a quartz substrate and the base liquid was allowed to evaporate. Magnetic hysteresis loops of SPIONs were measured at room temperature and at a low temperature of 6 K and are depicted in Figure 9. Field cooled (FC) and zero field cooled (ZFC) magnetization measurements were carried out from 6 K to 300 K at an applied field of 30 mT.

The room temperature M-H loop exhibits negligible coercivity (9 Oe) and remanence, which signifies the superparamagnetic nature of particles. This is further verified by fitting experimental curve with the modified Langevin function [27] for the particles with normal size distribution (Figure 9b).

$$M/M_s = L(a), \text{ where } a = mH/k_BT \qquad (1)$$

and

$$L(a) = \operatorname{Coth}(a) - 1/a \qquad (2)$$

with normal distribution of particles having a width "*b*" the function get modified to

$$L(a) = (1/2ba)\ln\frac{(1-b)\sinh(a(1+b))}{(1+b)\sinh(a(1-b))} \qquad (3)$$

where M is the magnetic moment for an applied field H, M_s is the spontaneous magnetization, k_B is the Boltzmann constant and T the temperature. At 6 K the coercivity is 125 Oe and the particles are in a thermodynamically blocked state. This is evident from ZFC measurement also. The saturation magnetization at room temperature, calculated by extrapolating the linear portion of magnetisation versus inverse of the applied field at higher field values is 0.418 emu/cc (Figure 9b (inset)). For a concentration of 6.288 mg/cc, the specific magnetization of the magnetic particles is 67emu/g. It is clearly seen that the magnetic moment is saturated at low applied fields and there is no further variation of moment even at high applied fields. If there are any traces of hematite, the moment even at high applied fields would not have been saturated. This is yet another evidence for the non occurrence of nonmagnetic iron oxide phase.

The specific magnetization of 67 emu/g for SPIONs is reasonably a good value and is sufficient for biomedical applications. The M (H) hysteresis loops of SiO_2-coated iron oxide nanoparticles taken at 300 and 10 K are depicted in Figure 10a. The silica-modified iron oxide nanoparticles exhibited negligibly small remanence as well as coercive field (H_C = 14 Oe) at room temperature (Figure 10b). This shows that the particle growth has not occurred during the process of silica modification.

The hysteresis loop opens at lower temperatures and coercivity increases to H_C = 122 Oe at 10 K. The latter, accompanied with an increase of the saturation magnetization, which is expected for the nanoparticles [28] as the thermal randomization is decreased.

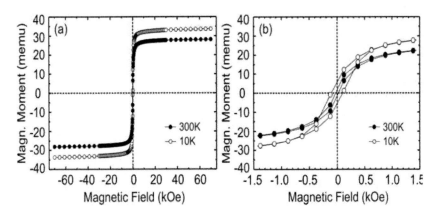

Figure 10. Hysteresis curves of silica coated SPIONs.

The effect of thermal activation and magnetic relaxation was studied by measuring zero-field and field cooling (ZFC/FC) curves. The ZFC and FC moments were measured at an applied field of 30mT (Figure 11a). The ZFC shows a broad blocking behaviour with a maximum moment at 140K, and above this temperature, it decreases gradually. This signifies the distribution of energy barriers present in the sample, and that they act as an ensemble of interacting fine particles. This may be collectively due to the randomly oriented surface spins, the size distribution [29, 30], and the inter-particle interactions. However, the presence of surfactant may eliminate the surface anisotropy as is reported by Roca et al. [31].

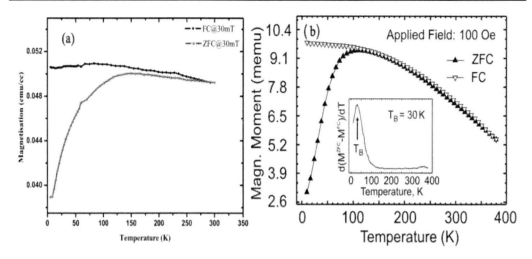

Figure 11. (a) Zero Field Cooled (ZFC) and Field Cooled (FC) moment variation at 30mT. 10 (b) ZFC/FC on Silica–SPIONs at 10mT.

The hydrodynamic length of citric acid is nearly 0.7nm that gives a physical separation of 1.4nm between particles' surfaces, since the base liquid has been dried off before subjecting to magnetization measurements. Thus there could exist among particles strong dipolar interaction that cause an increase in the effective energy and enhanced magnetic volume. This explains the increased value for the blocking temperature from the expected one. It's also seen that both the FC and ZFC curves show almost little decrease with temperature which establishes the inter particle interaction that competes the thermal fluctuation [32].

The interacting nature and hence the enhanced magnetic volume will spoil the performance in biomedical applications. The ZFC/FC measurements were carried out on powder samples of silica modified SPIONs, before dispersing them in PEG. For the FC study, the temperature changes from 300 down to 5 K in the presence of a small magnetic field. Figure 11b depicts the ZFC/FC measurements carried out on the silica-modified iron oxide nanoparticles at 100 Oe. The behavior of ZFC/FC curves is typical for the system of non interacting superparamagnetic particles, which is an essential criterion for their applicability in medical therapy.

These properties are not seen in ferrofluids prepared from CA-coated iron oxide nanoparticles (Figure 11a) indicating that the SiO_2 surface modification can successfully tune the magnetic properties as well as the inter-particle interactions. The ZFC/FC curves for SiO_2-coated iron oxide nanoparticles bifurcate at a temperature of about 115 K, which is close to the T_{max} on the ZFC (100 K). The latter suggests a rather narrow distribution of blocking temperatures of individual nanoparticles in assembly indicating a narrow particle size distribution [33, 34]. The blocking temperature is related to the distribution of energy barriers [35, 36] and can be estimated from the first derivative on temperature of the difference between the ZFC and FC moments (inset in Figure 11b). The maximum of the derivative corresponds to the blocking temperature of $T_B = 30$ K. This value is related to the nucleation volume, V, responsible for the magnetization reversal in magnetic nanoparticles. Assuming a thermally activated magnetization reversal with a switching rate following the Arrhenius law with a time constant τ_m (related to the measurement time, typically 100 s), the nucleation volume can be estimated using the expression $K \cdot V = k_B T_B \cdot \ln(\tau_m/\tau_0)$, with *K* being the anisotropy constant of Fe_3O_4 nanoparticles and $\tau_0 \approx 10^{-10}$ s an attempt frequency. The magnetic anisotropy constant of about $3x10^5$ erg/cm^3 was measured for Fe_3O_4 nanoparticles [37, 38]. Using the experimentally determined value for T_B, the nucleation volume of 345 nm^3 was found, which corresponds to the diameter of about 8.7 nm for a spherical nanoparticle. The latter is in a good agreement with the size of the iron oxide nanoparticles extracted from XRD and TEM data.

Theoretical Analysis for Magnetic Heating

Since the hysteresis is almost negligible as is concluded from M-H loop, the power dissipation is due to the relaxation processes. The power loss produced in an applied AC magnetic field as a function of the relaxation time is given by [39, 40]

$$P = \frac{(mH\omega\tau)^2}{2\tau V k_B T (1 + \omega^2 \tau^2)} \tag{5.4}$$

where m is the magnetic moment of the particle, V the particle volume, τ the relaxation time, H is the strength of the applied external magnetic field and ω is the angular frequency of the applied AC magnetic field. The relaxation is assumed to be arising from both Brownian and Neel relaxation mechanism and the total relaxation time τ is given by the equation

$$\frac{1}{\tau} = \frac{1}{\tau_N} + \frac{1}{\tau_B} \tag{5.5}$$

the Neel relaxation time τ_N is [40]

$$\tau_N = \frac{\sqrt{\pi}\tau_o \exp(KV / k_B T)}{2\sqrt{KV / k_B T}} \tag{5.6}$$

and the Brownian relaxation time τ_B is

$$\tau_B = \frac{4\pi\eta r^3}{k_B T} \tag{5.7}$$

where τ_o is the characteristic time constant $\sim 10^{-9}$ s , K the anisotropy constant, η is the viscosity of the medium, and r the hydrodynamic radius of the particle. Since the viscosity of living tissue is very high, the Brownian relaxation time becomes very large. So Neel relaxation dominates when the particle is functionalized and introduced for hyperthermia application in the living cells. So the power loss and hence the heat generated becomes a function of domain magnetization, anisotropy and volume of the particle for an AC field of fixed strength and frequency.

The anisotropy constant K is calculated from the relation

$$KV = 25\, k_B T_B \tag{5.8}$$

where T_B is the blocking temperature obtained from the ZFC measurements. The anisotropic constant calculated is of the same order as that of bulk iron oxide (1.1×10^5 erg/cc) [41] and is closer to the values reported in literature for iron oxide suspended in water [42, 43].

The power loss (also known as specific absorption rate SAR) of the prepared fluid particles, simulated as function of AC frequency in the range of 100 kHz to 900 kHz is plotted and is presented in Figure 12. The results obtained are consistent with the earlier calculations carried out by Okawa et al. [39], where the power loss for varied sizes is evaluated at a noninvasive frequency 120 kHz. Zhang et al. [44] has reported the SAR variation with particle size at still lower frequency of 55 kHz. The applied frequency for maximum power loss depends on the magnetic diameter where the Neel mechanism alone is considered. It is reported that the optimum size for non invasive frequencies lie around 12-14 nm [39]. Recent simulations [45] show that the power dissipation at an applied frequency of 800 kHz is 80 W/g and for 200 kHz, 10 W/g for an applied field strength of 200 Oe. In this study the corresponding values of power loss are 330 W/g and 20 W/g respectively. Li et al. [42] studied the variation of SAR with the viscosity of the fluid in which an increase in SAR with viscosity was reported till twice the viscosity of water. However at 55 kHz and 200 Oe, for water suspended fluids, they obtained a power loss of 57 W/g. It is seen from figure 12 that the power loss increases with frequency. The optimum frequency for required heat generation could be selected on the basis of the actual experimental condition where magnetic hyperthermia needs to be performed.

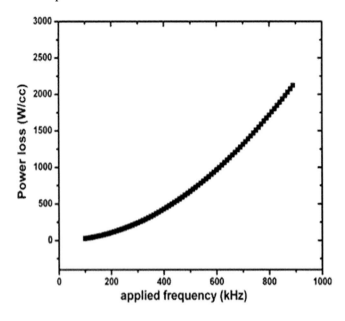

Figure 12. Power loss spectrum as a function of AC frequency for the ferrofluid.

Cytotoxicity of Ferrofluid Nanoparticles

Biocompatibility and non-toxicity are the prerequisites which determine the relevance of nanoparticles in *invitro* or *invivo* applications. Figure 13 shows the viability of HeLa cells in the presence of different concentrations of citric acid coated aqueous ferrofluid and silica-coated PEG-based ferrofluid nanoparticles. Here, we incubated HeLa cells with 0.1 - 1000 μg mL^{-1} concentrations of ferrofluid nanoparticles for 24 hours and the viability was measured as a function of mitochondrial dehydrogenase enzyme production using MTT assay. It is observed that the silica-coated nanoparticles were not toxic to HeLa cells in culture up to a concentration of 100 μg mL^{-1}.

However, the viability was decreased by ~ 25% when the concentration was further increased to 1000 μg mL^{-1}. MTT assay results for citric acid coated aqueous ferrofluid nanoparticles show that the viability decreased progressively with increasing concentration of nanoparticles and reached a minimum of ~40% viability with 1000 μg mL^{-1}. MTT assay is standard for quantifying the redox activity of mitochondrial dehydrogenase enzyme in living cells, and any decrease in the reduction of MTT is an index of mitochondrial damage and cell death. The cell viability test shows that the silica encapsulation enhances the compatibility, and the temperature dependant magnetic moment variation shows the reduction in inter-particle interaction due to surface modification with silica.

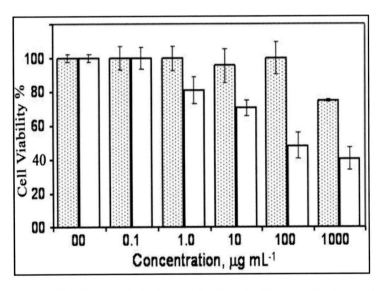

Figure 13. Histograms of MTT assays for HeLa cells incubated with different concentrations of citric acid coated aqueous ferrofluid (Bars with no pattern) and silica-coated PEG-based ferrofluid (Bars with dotted pattern). Values are given as percentage cell viability ± standard Error.

Optical Measurements

The synthesis of SiO$_2$-coated iron oxide ferrofluids open possibilities for synthesizing multifunctional ferrofluids having both magnetic as well as optical properties, such as photoluminescence. A number of researchers have reported the photoluminescence properties of silica nanoparticles, nanoshells and nanospheres [46, 47].

Attempts to observe photoluminescence on SiO$_2$-modified iron oxide ferrofluids were in vain. Pure silica nanospheres and nanoshells are supposed to be exhibiting luminescence at about 450 nm with an excitation wavelength of 360 nm.

The SPIONs have an optical band gap of ~2.65 eV (465 nm) as obtained from optical measurements. The diffused reflectance spectroscopy was employed to calculate the band gap of the Fe$_3$O$_4$ nanoparticles. The band gap was measured from tauc plots of $(\alpha h\nu)^2$ versus hν. The analysis shows that the direct band absorption takes place at energy of 2.65 eV (465 nm) (Figure 14), and is matching with the reported bandgap values of ultrafine Fe$_3$O$_4$ [48]. The obtained value is close to the luminescence line (450 nm) of the silica nanoshells [49].

This may cause the effective quenching of the luminescence of silica by iron oxide nanoparticles. Hence, the incorporation of SPIONs in the silica matrix might have resulted in the quenching of the luminescence of silica by absorbing this radiation as it is already observed in various heterostructures [50-52]. Time resolved measurements are required to probe the photoluminescence quenching properties of these core-shell heterostructured ferrofluids. The silica modification of the iron oxide nanoparticles opens the possibility for hybridizing their surface with various biocompatible dyes, and then these ferrofluids can exhibit multifunctional properties such as both magnetic and optical properties.

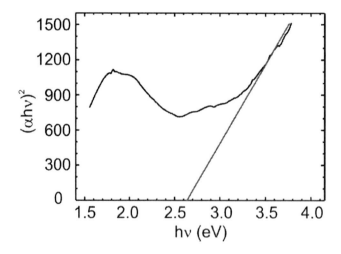

Figure 14. Band gap calculation for Fe$_3$O$_4$ nanoparticles using Tauc relation.

- He La cells: (Human Cervical Adeno carcinoma Cells).

It is the first immortal cell line and was employed in the development and testing if the polio vaccine. According to the genetic reports, it can be concluded that HeLa cells contains human papilloma virus 18 (HPV 18) sequences and express p53 (tumour suppression protein) at a low level. Levels of PRB (Retinoblastoma suppressor) are normals. HeLa cells are positive for the protein keratin and lysophosphatidylcholine (lyso-PC) which indices AP-1 activity and c-jun N-terminal kinase activity (JNK1) by a protein kinase C-independent pathway.

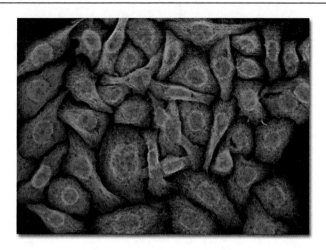

Courtesy.
http://www.google.com.br/imgres?imgurl=http://www.olympusfluoview.com/gallery/cells/hela/images/
helasmall.jpgandimgrefurl=http://www.olympusfluoview.com/gallery/cells/hela/helacells.htmlandh=26
0andw=358andsz=52andtbnid=gBZcXm9uUdLj9M:andtbnh=88andtbnw=121andprev=/search%3Fq%
3DHeLa%2Bcells%26tbm%3Disch%26tbo%3Duandzoom=1andq=HeLa+cellsandhl=pt-
BRandusg=__m4AhjPyfR4pb58--l391tioHdIo=andsa=Xandei=WSEQT4Frhs7hBK29xNYD
andved=0CCIQ9QEwBA.

CONCLUSION

Highly stable aqueous ferrofluid of SPIONs with citric acid as surfactant has been synthesized by controlled chemical co-precipitation method. The structural investigation by XRD and TEM show good toning in respect of the particle size of 9.5 nm. The narrow size distribution obtained is highly desirable for biomedical applications, since the relaxation time varies with the size, and distribution in size will lead to distribution of relaxation time. The magnetic analysis shows that the nanoparticles are superparamagnetic in nature and hence the coercivity is negligible.

The saturation magnetization value (67 emu/g for the water based fluid) is found to be suitable for various biomedical applications especially for magneto hyperthermia. Based on the magnetic measurements, the power dissipation in an alternating magnetic field of 500 gauss as a function of applied frequency is calculated. The results reveal the possibility of employing these nanoparticles in magnetic hyperthermia. In order to enhance the cell compatibility and to reduce the inter particle interaction, the SPIONs were surface modified with silica by Stober process. The analysis shows the particles are embedded in silica and are magnetically non interacting and hence good from a biomedical application point of view.

REFERENCES

[1] Catherine C. Berry, *J. Mater. Chem.*15, 543–547 (2005).
[2] D. K. Kim, M. Mikhaylova, F. H. Wang, J. Kehr, B. Bjelke, Y. Zhang, T. Tsakalakos, M. Muhammed, *Chem. Mater.* 15, 4343 (2003).

[3] N. A. Brusnetsov, V. V. Gogosov, T. N. Brusnetsova, A. V. Sergeev, N. Y. Jurchenko, A. Kuznetsov, O. A. Kuznetsov, L. I. Shumakov, *J. Magn. Magn. Mater.* 225, 113 (2001).

[4] D. L. Holligan, G. T. Gillies, J. P. Dailey, *Nanotechnology* 14, 661 (2003).

[5] C. C. Berry, A. S. G. Curtis, *J. Phys. D: Appl. Phys.* 36, R198 (2003).

[6] M. Babincova, V. Altanerova, C. Altaner, C. Bergeman, P. Babinec, *IEEE Trans. Nanobioscience*, 7, 15 (2008).

[7] T. Y. Liu, S. H. Hu, T. Y. Liu, D. Liu, S. Y. Chen, *Langmuir* 22, 5974 (2006).

[8] Lutz J F, Stiller S, Hoth A, Kaufner L, Pison U and Cartier R, *Biomacromolecules*, 7, 3132 (2006).

[9] C. Scherer, Figero A. M. Neto, *Braz. J. Phys.* 35 3, 718 (2005).

[10] Y. Bao, A. B. Pakhomov, and Kannan M. Krishnan, *J APpl.Phys.* 99 08H107(2006).

[11] P. Tartaj, M. P. Morales, S. Veintemillas-Verdaguer, T. Gonzalez-Carreno, C. J. Serna *J. Phys. D: Appl. Phys.* 36 , R182 (2003).

[12] S.-H. Chung, A. Hoffmann, K. Guslienk, S. D. Bader, C. Li2, B. Kay, L. Makowski, and L. Chen *J. Appl.Phys.* 97 10R101 (2005).

[13] Petri-Fink, H. Hofmann, IEEE Trans. *Nanobioscience* 6 (4) , 289 (2007).

[14] H. Shao, T-J Yoon, M. Liong, R. Weissleder and H. Lee, *Beilstein J. Nanotechno.* 1 142 (2010).

[15] Q. A. Pankhurst, J. Connolly, S. K. Jones, J. Dobson, *J. Phys. D: Appl. Phys.* 36, R167 (2003).

[16] R. Hiergeist, W. Andrä, N. Buske, R. Hergt, I. Hilger, U. Richter, W. Kaiser, *J Magn.Magn.Mater.* 201, 420 (1999).

[17] J. Giri, P. Pradhan, T. Sriharsha, D. Bahadur, *J. Appl. Phys*, 97, 10Q916 (2005)

[18] R. E. Rosensweig, *Ferrohydrodynamics*, Cambridge University Press, New York, 1985

[19] E. Dubois, V. Cabuil, F. Boue, R. Perzynski, *J. Chem. Phys.* 111 , 7147 (2001).

[20] J. de Vincente, A. V. Delgado, R. C. Plaza, J. D. G. Duran, F. Gonsalez-Caballero, *Langmuir* 16, 7954 (2000).

[21] L. F. Pavon, O. K. Okamoto, *Einstein* 5, 74 (2007).

[22] D. K. Kim, M. Mikhaylova, Y. Zhang and M. Muhammed, *Chem. Mater.* 15, 1617 (2003).

[23] Ajay Kumar Guptaa and, Mona Gupta, *Biomaterials* 26 (2005) 3995–4021.

[24] Z. Krpeti, F. Porta, E. Caneva, V. D. Santo, G. Scarı, , *Langmuir*, 26(18) 14799–14805 (2010).

[25] Xiao-Dong Wanga, Zheng-Xiang Shena, Tian Sanga, Xin-Bin Chenga, Ming-Fang Lib, Ling-Yan Chena and Zhan-Shan Wanga, *Journal of Colloid and Interface Science* 341, 23-29 (2010).

[26] R. M. Silverstein and G. C. Bassler, *Spectrometric identification of organic compounds, John Wiley and Sons, Inc.,* New York (1963).

[27] T. N. Narayanan, A. P. Reena Mary, M. M. Shajumon, L. Ci, P. M. Ajayan, M. R. Anantharaman, *Nanotechnology* 20, 055607 (2009).

[28] B. D. Cullity, *Introduction to Magnetic materials, Addison-Wesley Series Publishing Company, Inc.* Philippines (1972).

[29] S. J. Lee, J. R. Jeong, S. C. Shin, J. C. Kim, J. D. Kim, *J. Magn. Magn. Mater.* 282 , 147(2004).

[30] X. Battle, A. Labarta, *J. Phys. D: Appl. Phys.* 35, R15 (2002).

[31] G. Roca, M. P. Morales, K. O'Grandy, C. J. Serna, *Nanotechnology* 17, 2783, (2006).

[32] J. Dai, J. Q. Wang, C. Sangregorio, J. Fang, E. Carpenter, J. Tang, *J. Appl. Phys.* 87, 7397, (2000).

[33] P. A. Joy and S. K. Date, *J. Magn. Magn. Mater.* 222, 33 (2000).

[34] R. W. Chantrell, N. S. Walmsley, J. Gore and M. Maylin, *J. Appl. Phys.* 85, 4340 (1999).

[35] J. C. Denarding, A. L. Brandl, M. Knobel, P. Panissod, A. B. Pakhomov, H. Liu andX. X. Zhang, *Phys. Rev. B* 65, 064422 (2002).

[36] J. M. Vargas, J. Gómez, R. D. Zysler and A. Butera, *Nanotechnology* 18, 115714 (2007).

[37] J. M. Vargas, E. Lima Jr, R. D. Zysler, J. G. S. Duque, E. De Biasi and M. Knobel, *Eur. Phys. J. B* 64, 211 (2008).

[38] D. Arelaroa, E. Lima Jr, L. M. Rossi, P. K. Kiyohara and H. R. Rechenberg, *J. Magn. Magn. Mater.* 320, 335 (2008).

[39] K. Okawa, M. Sekine, M. Maeda, M. Tada, M. Abe, N. Matsushida, K. Nishio, H. Handa, *J. Appl. Phys.* 99, 08H102 (2006)

[40] R. E. Rosenswieg, *J. Magn. Magn. Mater*, 252, 370, (2002).

[41] B. D. Cullity, Introduction to Magnetic materials Philippines: Addison Wesley, London, 1972.

[42] Y. Z, Li, -C. G Hong._ M. W, Xu, *J. Magn. Magn. Mater.* 311 ,228–233, (2007).

[43] M. Ma, Y. Wu, J. Zhou, Y. Sun, Y. Zhang and N. Gu, *J. Magn. Magn. Mater.* 268,33(2004).

[44] L. Y. Zhang, Y. H. Dou, L. Zhang and H. C. Gu, *Chin. Phys. Lett.* 24, 483, (2007).

[45] S. Purushotham and R.V. Ramanujan, *J. Appl. Phys.* 107, 114701, (2010).

[46] S. Thomas, D. Sakthi Kumar, Y. Yoshida and M . R. Anantharaman, *J. Nanoparticle Res.* 10, 203 (2008).

[47] T. N. Narayanan, D. Sakthi Kumar, Y. Yoshida, M. R. Anantharaman, *Bull. Mater. Sci.* 31, 759 (2008).

[48] S. S Nair, M. Mathews and M. R. Anantharaman, *Chem. Phys. Lett.,* 406, 398 (2005).

[49] Y. D. Glinka, S. H.Lin and Y. T. Chen, *Phys. Rev. B,* 66, 035404 (2002).

[50] X. Li, J. Qian, L. Jiang and S. He, *Appl. Phys. Lett.* 94, 063111 (2009).

[51] K. Mazunari, Y. Ito and Y. Kanemitsu, *Appl. Phys. Lett.* 92, 211911 (2008).

[52] H. Li and L. Rothberg, *PNAS* 101, 14036 (2004).

In: Ferrofluids
Editors: Franco F. Orsucci and Nicoletta Sala

ISBN: 978-1-62808-410-8
© 2013 Nova Science Publishers, Inc.

Chapter 7

INDUSTRIAL APPLICATIONS OF FERROFLUIDS

*Swapna Nair**
Departmento de Engenharia Cerâmica e do Vidro and CICECO,
Universidade de Aveiro, Aveiro, Portugal

ABSTRACT

Ferrofluids are widely employed in versatile applications like in sealing, as loud speaker coolant, lubricants, in pressure sensors, in display devices, in biomedical applications like targeted drug delivery, hyperthermia, as a low side effect MRI contrasting agent, as magnetically tunable optical filters, as tunable diffraction gratings, as magnetic field and moment sensor etc [1-11]. This chapter describes these applications of ferrofluids. Optical, non linear optical and biomedical applications of ferrofluids are elaborated in detail in the previous chapters. Hence this chapter mainly concentrates on the industrial application of ferrofluids.

Ferrofluids are widely employed in industry. Ferrofluid filled loud speakers are already in markets as dampers. Laser head in a CD or DVD player most probably has a drop of ferrofluid as a damper. They are widely used as different type of sensors. Ferrofluids used as seals and coolants. Each of these applications are detailed in the following sections.

---

* E-mail: swapna.s.nair@gmail.com.

FERROFLUID SEALS

One of the most important industrial applications of ferrofluids is their employment as fluid seals [1-11]. They are used extensively to form liquid seals around the spinning drive shafts in hard disks. The rotating shaft is surrounded by magnets. A small amount of ferrofluid, placed in the gap between the magnet and the shaft, will be held in place by its attraction to the magnet. The fluid of magnetic particles forms a barrier which prevents debris from entering the interior of the hard drive. The most important advantage is that the amount of liquid seal needed is very less as a tiny drop of magnetic fluid will go on rotating with the magnetic field.

Ferrofluid based rotary seals operate with no maintenance and extremely low leakage in a very wide range of applications. They are used widely in variety of industrial and scientific applications. Generally, they are packaged in mechanical seal assemblies called rotary feedthroughs. Rotary feedthroughs consist of mainly 3 parts,

1. A central shaft
2. Ball bearings
3. An outer housing

The ball bearings are used to maintain the shaft within the seal gap as well as to support external loads. The bearings are the only mechanical wear-items, as the dynamic seal is actually a series of ferrofluid rings made of ultra-low vapour pressure, oil-based liquid held magnetically between the rotor and stator.

This enhances the operating life and equipment maintenance cycles and the drag torque is maintained very low. The magnet material is permanently charged and requires no electrical power or other re-energizing or maintenance. Ferrofluid-sealed feedthroughs reach performance levels that other technologies can't achieve, by optimizing features such as ferrofluid viscosity and magnetic strength, magnet and steel materials, bearing arrangements, and water cooling for applications with extremely high speeds or temperatures. Ferrofluid-sealed feedthroughs routinely operate in environments including ultra-high vacuum (below 10^{-8} mbar), temperatures over 1,000 °C, tens of thousands of RPM, and multiple-atmosphere pressures.

Magnetic liquid seals are engineered for a wide range of applications and exposure but are generally limited to sealing gases and vapours, not direct pressurized liquid. Each particular combination of construction materials and design features has practical limits with respect to temperature, differential pressure, speed, applied loads and operating environment; however, these limits can generally be overcome by carefully selecting or designing the device for the application.

Necessary features may include multiple ferrofluid stages, water cooling, customized materials including metals, permanent magnets and ferrofluid, and exotic bearings. Ferrofluid-based seals have extremely low leak rates - almost immeasurable with laboratory equipment - however they cannot reach the levels of welded connections or other all-metal, static (non-rotating) seals. Numerous research work have been conducted in designing such seals and investigating and tailoring of their properties [12-15].

AS A LUBRICANT AND HEAT TRANSFER AGENT:

Lubricants are the materials which are used to reduce the wear and tear caused by the friction. Ferrofluids are often employed as lubricants because of their fluidic nature with the added advantage of being controllable in their flow with the application of an external magnet [1, 3]. Hence magnet can glide across smooth surfaces with minimal resistance if ferrofluids are employed as lubricants.

An external magnetic field imposed on a ferrofluid with varying susceptibility (e.g., because of a temperature gradient) results in a nonuniform magnetic body force, which leads to a form of heat transfer called thermomagnetic convection. This form of heat transfer can be useful when conventional convection heat transfer is inadequate; e.g., in miniature microscale devices or under reduced gravity conditions.

In Loud Speakers

Ferrofluids are commonly used in loudspeakers as a heat transfer agent and as a damper which passively damp the speaker cone movement [1-3]. They reside in what would normally be the air gap around the voice coil, held in place by the speaker's magnet. Since ferrofluids are superparamagnetic, they obey Curie's law, thus become less magnetic at higher temperatures. A strong magnet placed near the voice coil (which produces heat) will attract cold ferrofluid more than hot ferrofluid thus forcing the heated ferrofluid away from the electric voice coil and toward a heat sink. This is an efficient cooling method which requires no additional energy input. [6] Ferrofluid based speakers are already in markets.

Audio ferrofluids are based on two classes of carrier liquid: synthetic hydrocarbons and esters. Both oils possess very low volatility and high thermal stability. The choice of fluid is dictated by the environmental considerations of the application (e.g. humidity, adhesives, contact with water, solvent vapors and reactive gases) combined with the best balance of magnetization and viscosity values to optimize the acoustical performance.

By varying the quantity of magnetic material in a ferrofluid, and by using different carrier liquids, it can be tailored to meet a variety of needs. The saturation magnetization (the maximum value of the magnetic moment per unit volume when all the domains are aligned) is determined by the nature of the suspended magnetic material and by the volumetric loading of the material. The physical and chemical properties such as density and viscosity correspond closely to those of the carrier liquid.

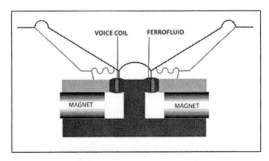

As Sensors

Different sensors used to sense pressure, volume, temperature magnetic field etc is fabricated Magnetic fluid by employing ferrofluids [16]. Bacri et al. designed a pressure sensor based on ferrofluids. [17]. Stanci et al. proposed the design and fabrication of ferrofluids for volume sensing applications. [18]. Temperature sensors based on magnetic fluids, have already designed [19]. A flow transducer for cold water using ferro fluids was proposed by Monica et al. [20, 21]. Popa et al. studied in detail regarding the application of ferrofluids in inductive transducers and aerodynamic measuring devices [22-24]. It is reported recently that magnetic fluid films can be employed for sensing magnetic field and moment [25].

Bio Medical Applications of Ferrofluids

Ferrofluids can be employed in different biomedical applications like targeted drug delivery, hyperthermia etc [26-28]. In medicine, ferrofluids are used as contrast agents for magnetic resonance imaging and can be used for cancer detection [29]. The ferrofluids are in this case composed of iron oxide nanoparticles and called SPION, for "Superparamagnetic Iron Oxide Nanoparticles"

There is also much experimentation with the use of ferrofluids in an experimental cancer treatment called magnetic hyperthermia. It is based on the fact that a ferrofluid placed in an alternating magnetic field releases heat.

Tunable Filters, Defect Sensors and Intelligent Cooling

Recently, it has been demonstrated that ferrofluids of suitable composition can exhibit extremely large enhancement in thermal conductivity(k) (i.e. ~ 300% of the base fluid thermal conductivity). Studies confirmed that the large enhancement in k is due to the efficient transport of heat through percolating nanoparticle paths. Special magnetic nanofluids with tunable thermal conductivity to viscosity ratio can be used as multifunctional 'smart materials' that can remove heat and also arrest vibrations (damper). Such fluids offer exciting applications in microfluidic devices, micro and nano electromechanical systems (MEMS and NEMS) and other nanotechnology based miniature devices [30, 31].

Highly monodispersed magnetic emulsions with droplets change the colour upon application of external magnetic field due to Bragg reflection from one-dimensional linear array of droplets. By taking advantage of the ordering property of the ferrofluid, a new application for this fluid to detect defects in ferromagnetic materials and components based on magnetic flux leakage (MFL) technique has been developed. [32] Optical filters are used to select different wavelengths of light. A filter works by excluding all but a limited set of wavelengths. Color imaging systems benefit from the use of precision optical filters, which control the spectral properties of light and color separation to exacting tolerances. In addition, system performance improves with the elimination of wavelengths outside the visible spectrum. In spectroscopic and interferometric experiments, involving different wavelengths of light, different types of interference filters are required to eliminate stray light entering the detector head. In those experiments, filters are to be replaced, whenever the light source (wavelength) is changed. e.g.: when a He-Ne laser is replaced by an Ar-ion laser. The replacement of filters is cumbersome, especially when the wavelength is changed continuously with tunable type of lasers. In order to avoid the problems mentioned above, researchers at Indira Gandhi Centre for Atomic Research (IGCAR, India) has developed a novel tunable optical filter using ferrofluid emulsion [33]. The filter comprises of ferrofluid-based emulsion cell, a miniature solenoid and a variable direct current source for changing the magnetic field. Here, by varying the magnetic field, we tune and select the desired wavelength.

The main advantages of the new optical filter are (a) a single filter can be used for a range of central wavelengths, where the desired central wavelength region can be tuned by external magnetic field (b) it is suitable for selecting wavelengths in the ultraviolet, visible and infrared regions (c) there is no need for changing the optical element for different wavelength regions (d) tuning can be easily achieved by changing the field strength (e) The spectral distribution can be controlled by adjusting the polydispersity of the emulsion (f) The intensity of the transmitted light can be controlled by changing the emulsion concentration (g) it is simple to operate and less expensive compared to the existing filters.

Optical and Non Linear Optical Applications

Ferrofluids may be used as a dynamic mask for photolithography and optical imaging applications such as video projection. This device allows the user to tailor a pattern for photolithography using the ferrofluidic deformable mirror to reflect light in a controlled pattern onto another surface.

Unlike the generally used masks for photolithography, the ferrofluid based device can be tuned magnetically for their transparency so that different patterns can be produced from the single device, eliminating the need for multiple masks and potentially saving time and cost. For projection, ferrofluidic deformable mirrors may be used, as an alternative to Dynamic Light Processing (DLP) technologies, which are currently used in high definition video.

Brousseau et al. reported a ferrofluid mirror design which will result in an inexpensive adaptive optics element with large stroke for use in ophthalmic imaging [34].

Deformable mirrors based on ferrofluids were designed and the properties were studied already both experimentally and theoretically which have applications in astronomy and astrophysics especially in liquid mirror telescope (Rayleigh scatterer), as a basic for lunar telescope etc.. It is also proposed for retinal imaging adaptive optics systems. [35-38]

Non linear optical applications of ferrofluids are recently investigated and it is concluded that thin ferrofluid films can be employed as an optical limiter [39]. Optical limiters are the devices which allows low to moderate intensities of light while block the high intensity lights (non linear intensities). An ideal optical limiter possesses more than 90% light transmission in the linear regime while block the non linear intensities almost completely. Details are presented in chapter.

Art

Some art and science museums have special devices on display that use magnets to make ferrofluids move around specially shaped surfaces in a fountain show-like fashion to entertain guests. Sachiko Kodama is known for her ferrofluid art. The Australian electronic rock band, Pendulum, used ferrofluid for the music video for the track, Watercolour. The design house Krafted London was responsible for the ferrofluid FX in the video. The post-metal band Isis also uses a Ferrofluid in the music-video for 20 Minutes/40 Years.

CZFerro, an American art studio, began using ferrofluid in its productions in 2008. The works consist of ferrofluid displayed in a unique suspension solution. These works are often used as conversation pieces for offices and homes.

Courtesy: google images: vimeo.com, 2192851955_e03dd2efd3.jpg.

Courtesy : http://nanobioart.com/nanolab/recent-posts/page/34/

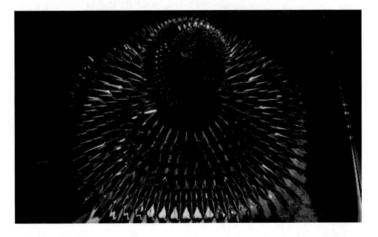

Courtesy: http://www.msichicago.org/whats-here/exhibits/science-storms/the-exhibit/atoms/ferrofluid/

CONCLUSION

Industrial applications of ferrofluids are enormous and the applications detailed in this chapter cover only some of them. More researchers are now working in unravelling new applications and developing new devices based on ferrofluids.

REFERENCES

[1] R.E. Rosensweig, *Ferrohydrodynamics*, Cambridge Univ. Press, pp.344, 1985.
[2] J.L. Neuringer, R.E. Rosensweig, Phys. Fluids 7(1964)1927.

[3] B. Berkovsky, V. Bashtovoi (Eds.), *Magnetic fluids and applications handbook,* Begell House, New York, pp.831, 1996.

[4] S. Odenbach (Editor), *Ferrofluids: Magnetically controllable fluids and their applications,* Lecture, Notes in Physics, Springer-Verlag, 253 pags (2002).

[5] S. W. Charles, The preparation of magnetic fluids, in: S. Odenbach (Editor), *Ferrofluids: Magnetically controllable fluids and their applications,* Lecture Notes in Physics, Springer-Verlag, pp.3-18,2002.

[6] S. W. Charles, Preparation and magnetic properties of magnetic fluids, *Rom. Repts. Phys.,* 47, (1995) 249.

[7] L. Vekas, D. Bica, M. V. Avdeev, Magnetic nanoparticles and concentrated magnetic nanofluids: synthesis, properties and some applications, *China Particuology,* 5, 43 (2007).

[8] Anton, I.De Sabata, L. Vekas, Application orientated researches on magnetic fluids, *J. Magn. Magn. Mater.,* 85, (1990) 219.

[9] K. Raj, Magnetic fluids and devices: a commercial survey, in: B. Berkovsky, V. Bashtovoi (Eds.), *Magnetic fluids and applications handbook,* Begell House, New York, pp.657-751 (1996).

[10] V. Cabuil, J.C. Bacri, R. Perzynsky, YU. Raikher, Colloidal stability of magnetic fluids, in: B. Berkovsky, V. Bashtovoi (Eds.), *Magnetic fluids and applications handbook,* Begell House, New York, pp.33-56 (1996).

[11] M.I. Shliomis, Ferrohydrodynamics: Retrospective and Issues, in: S. ODENBACH (Editor), *Ferrofluids: Magnetically controllable fluids and their applications,* Lecture Notes in Physics, Springer-Verlag, pp.85-110, 2002.

[12] Tarapov, *Magnetohydrodynamics,* 8, (1972), 444.

[13] R. C. Shah and M. Bhat, *TribologyInternational,* 37, (2004) 441.

[14] Q. Zhang, S. Chen, S.Winoto, and E. Ong, "Design of high-speed magnetic fluid bearing spindle motor," *IEEE Trans. Magn.,* 37, (2001) 2647.

[15] P. Kuzhir, *TribologyInternational,* 41, (2008) 256.

[16] C.Cotae, O.Baltag, R.Olaru, D.Calarasu, D.Costandache. *Sens. and Actu.* 84, (2000) 246.

[17] J.C. Bacri, J. Lenglet, R. Perzynski, J. Servais, *J. Magn.Magn. Mater.*122, (1993), 399.

[18] Stanci, V. lusan, C. D. Buioca. *J. Magn.Magn. Mater,* 201, (1999) 394.

[19] http://www.nplindia.org/ferrofluid-based-temperature-sensor.

[20] M.S. Crainic, Z. Schlett. *J. Magn.Magn. Mater* 268, (2004), 8.

[21] M.S. Crainic, M. Cornel, D. Hie, Flow Meas. and Instru., 11, (2000) 101.

[22] N.C.Popa, A. Siblini, L.Jorat, *J. Magn.Magn. Mater* 242, (2002) 1240.

[23] N.C.Popa, I. De Sabata, Sens. and Actua. A 59 (1997), 197.

[24] N.C.Popa, I De Sabata, I.Anton, I.Potencz, L.Vekas, *J. Magn.Magn. Mater* 201, (1999) 385.

[25] S.S. Nair, S. Rajesh, V. S. Abraham and M. R. Anantharaman, *Bull.Mater.Scie,* 34, (2011), 245.

[26] T. Neuberger, B. Schopf, H. Hofmann, , M. Hofmann, B. Rechenberg,. *J.Magn. Magn. Mater.,* 293 (2005) 483.

[27] Q.A.Pankhurst, J.Connolly, S.K. Jones, J. Dobson, *J. Phys. D: Appl.Phys.,* 36(2003) R167.

[28] A.P. Reena Mary , T.N. Narayanan , V. Sunny , D. Sakthikumar , Y. Yoshida , P.A. Joy, M.R. Anantharaman , Nanosc. Res. Lett. 5, (2010), 1706.

[29] F. Maria Casula, P. Floris, C. Innocenti, A. Lascialfari, M. Marinone, M. Corti, A. Ralph Sperling, J. W. Parak , C. Sangregorio, *Chem. Mater.*, *22* (2010) 1739.

[30] P. D. Shima and John Philip, *J. Phys. Chem. C.*, 115, (2011) 20097.

[31] J. Philip, P. D. Shima, B. Raj, Appl. Phys. Lett. **92**, (2008)043108.

[32] J., C Babu Rao, B. Raj and T. Jayakumar, *Meas. Sci. Technol.* 10, (1999) N71.

[33] John Philip, T Jaykumar, P Kalyanasundaram. Baldev Raj, *Meas. Sci. Technol.* 14 (2003) 1289.

[34] D. Brousseau, E. F. Borra, A. M. Ritcey, M. C. Campbell, S. Thibault, J. Drapeau, and A. Naderian, "A New Ferrofluid Mirror for Vision Science Applications," in *Frontiers in Optics*, OSA Technical Digest (CD) (Optical Society of America, 2009), paper JWF2.

[35] P. Laird, E. F.Borra, R.Bergamesco, J.Gingras, L Truong, A. M. Ritcey, *Proc. SPIE* 5490, (2004) 1493.

[36] D.Brousseau, , E. F.Borra, S.Thibault, *Opt.Express* 15, (2007) 18190.

[37] Iqbal, and F. B.Amara, *Inter. J. Optomechatronics* 1, (2007) 180.

[38] E. F. Borra, D. Brousseau, A. Vincent, *Astron. Astrophys.* 446, (2006) 389.

[39] S. S. Nair, J. Thomas, C. S. Suchand Sandeep, M. R. Anantharaman, R.Philip, *Appl. Phys. Lett.* 92, (2008) 171908.

See also

http://www.directindustry.com/prod/ferrolabs/ferrofluid-based-seals-30165-136078.html
http://machinedesign.com/article/magnetic-fluids-tackle-tough-sealing-jobs-0217
http://liquidsresearch.com/en-GB/for_sealing_applications-52.aspx
http://www.sdmaterials.com/ferrofluid_seals.html
http://www.ferrotec.com/products/ferrofluidic/
http://www.directindustry.com/prod/ferrolabs/ferrofluid-based-seals-30165-136078.html

In: Ferrofluids
Editors: Franco F. Orsucci and Nicoletta Sala

ISBN: 978-1-62808-410-8
© 2013 Nova Science Publishers, Inc.

Chapter 8

MICROWAVE PROPERTIES OF FERROFLUIDS

K. Sudheendran[1] and Swapna Nair[2,]*

[1]Department of Physics, Sree Kerala Varma College, Thrissur, India
[2]Departmento de Engenharia Cerâmica e do Vidro and CICECO,
Universidade de Aveiro, Aveiro, Portugal

ABSTRACT

Microwaves (300 M Hz – 300 G Hz) have become very important for today's human life and which are extensively used in today's civilian and military communication systems as well as in domestic and industrial appliances. Due to the extensive expansion of the wireless communication networks, the microwave instruments have become part of the day to day life. At the same time it is important to note that microwave radiations are capable of producing harmful effects to the human body organs if exposed for a considerable time. Because of this, microwave absorbers are gaining importance in controlling the adverse effects of microwave. New designs for microwave absorbers are being predicted with the use of nano particulate inclusions in a liquid matrix. Ferrofluids are ideal materials in this direction. This chapter describes the dielectric magnetic and microwave absorption characteristics of certain ferrofluids. These results suggest that ferrofluids can be potentially integrated in to microwave absorbers.

Keywords: Ferrofluids, permittivity, microwave absorber

INTRODUCTION

Recently the radio Frequency radiations originating from the wireless communication systems has become a subject of concern because of it possible threats to the environment and the living organisms. The interaction of biological matter with an electromagnetic field is debated to be a possible cause for the formation of cancerous tissues. Even though there are no conclusive evidences on the ill effects of electromagnetic field radiation, scientist has

---

* E-mail: swapna.s.nair@gmail.com.

urged a caution for continues exposure to the electromagnetic fields. This has leads to the development of electromagnetic shielding technology which will provide an adequate way to shield or absorbs the electromagnetic fields.

Microwave absorbing materials for military applications have been investigated since the advent of radar systems. The design of a good absorber becomes very difficult since precise control over some magnetic and dielectric properties is often necessitated, and conventional materials become challenging or even impossible to integrate into the design.

In the mean time new designs for microwave absorbers are being predicted with the use of nano particulate incisions in a liquid matrix. In a colloidal RF absorber, the absorbance cross-section can be tailored through changes in size, composition, morphology and density of the nanoparticle inclusions [1].

These material properties can be accurately controlled through the particle synthesis process. Furthermore, the absorbance characteristics of the colloid can be tailored via choice of an appropriate liquid matrix in which the inclusions are suspended. With the choice of a liquid matrix containing unbound inclusions, the feasibility of modulation of RF absorption becomes possible without the use of mechanical or electrostatic actuation. Instead, absorption changes can be realized through application of chemically selective surface ligands such as bio molecules to the nanoparticle surfaces. These site-specific ligands can dictate particle aggregation and dissolution in the colloid, thus modulating the local absorption characteristics through changes in the particle volume fraction and the magnetic properties of the colloid.

Ferro fluids are stable colloidal suspensions of nonmagnetic particles usually having single domain super paramagnetic characteristics [2-5]. Stability against agglomeration and settling has been taken care during the synthesis technique. They can act as a system of non-interacting magnetic domains. They find extensive applications in versatile fields like in rotary sealing, pressure sensors, loud speaker coolants, display devices etc. [6, 7]. In the absence of an external magnetic field they exhibit usual properties of a stable colloid while in the presence of an applied magnetic field, their magnetic, optical and electrical properties could be greatly modified which throws open a new area in nanomaterials whose underlying Physics has not been investigated in detail. Ferrites are already known for their microwave absorption in UHF and VHF band. As evident from theory, the most important property that governs electromagnetic absorption is the dielectric permittivity and magnetic permeability at the desired bandwidth operation. Hence the ferrofluids can serve as an active material for microwave absorption

This chapter deals with the microwave and dielectric characteristics of magnetite based ferrofluids which helps to evaluate the applicability of these systems for the development of next generation microwave absorbers.

THEORY OF MICROWAVE ABSORPTION

Microwaves constitute only a small portion of the electromagnetic spectrum, but the applications have become increasingly important to investigate the material properties. Measurement of material properties like permeability, permittivity, conductivity etc will serve as a tool for investigating the intermolecular and intra molecular effects.

Bethe and Schwinger [8] suggested the cavity perturbation theory for the first time. According to them, the perturbation was caused by the insertion of small dielectric sample in to the cavity and by a small deformation of the boundary surface of the cavity.

By quantifying the perturbation occurred, various material properties can be measured. The fundamental idea is that the change in the overall configuration of the electromagnetic fields upon the introduction of the sample must be low. Based on this assumption, detailed derivation of the perturbation equation for the frequency shift upon introduction of a sample into a cavity was given by Waldron and Harrington.

The permittivity and permeability of a material in general is a complex quantity such as $\varepsilon_r = \varepsilon_r' - j\varepsilon_r''$ and $\mu = \mu' - j\mu''$.

The real and imaginary part of the permittivity are obtained from the perturbation equation for the frequency shift as,

$$\varepsilon_r' + 1 = \frac{(f_0 - f_s)}{2f_s} \frac{V_c}{V_s}$$

$$\varepsilon_r'' = \frac{V_c}{4V_s} \left(\frac{1}{Q_s} - \frac{1}{Q_0} \right)$$

Here Q_0 and f_0 represent the quality factor and resonance frequency of the cavity in the unperturbed condition respectively. and f_s are the corresponding parameters of the cavity loaded with the sample.

To study the dielectric properties of liquid, a container is required. For this purpose, a capillary tube of low- loss, fused silica is used.

If the frequency shift is measured from the resonance frequency f_t of the cavity loaded with an empty capillary tube rather than that with an empty cavity alone, the above equation becomes

$$\varepsilon_r' + 1 = \frac{(f_t - f_s)V_c}{2f_s} \frac{V_c}{V_s}$$

$$\varepsilon_r'' = \frac{V_c}{4V_s} \left(\frac{1}{Q_s} - \frac{1}{Q_t} \right)$$

Here, represent the quality factor of empty tube.
Similarly, the real and part of the complex permeability is given by,

$$\mu_r' + 1 = \left(\frac{\lambda_g^2 + 4a^2}{8a^2} \right) \frac{(f_t - f_s)}{f_s} \frac{V_c}{V_s}$$

where f_s and f_t are the resonant frequencies of the cavity with the sample and with the empty tube, respectively and a represent the λ_g is the guided wavelength and is given by

$$\lambda_g = \frac{2d}{p} \text{ where } p = 1,2,3,......$$

Reflection Coefficient of a Microwave Absorber

To qualitatively assess the potential of a microwave absorber material one has to quantify its reflection coefficient at different frequencies. A model of a simple one-layer electromagnetic absorber can be analyzed for this purpose as described in Figure 1. Such an absorber is constructed by one layer of the absorbing material (here ferrofluids) taken in a appropriate container, backed with a metal layer. Reflection coefficient of this structure for the normally incident EM waves can be calculated as follows

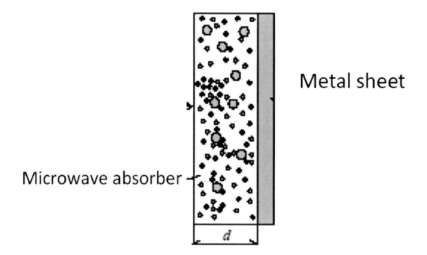

Figure 0. Design of a single layer microwave absorber.

For a microwave-absorbing layer backed by a metal plate as shown in figure 0, the normalized input impedance (Zin) at the absorber surface is given by

$$Z_{in} = \sqrt{\frac{\mu_r}{\varepsilon_r}} \tanh\left(j \frac{2\pi}{c} \sqrt{\mu_r \varepsilon_r} f \cdot d \right)$$

where μr is the complex permeability and ε_r is the complex permittivity and c is the velocity of electromagnetic waves in free space, f is the frequency, and d is the thickness of the absorber. The reflection coefficient (dB) is a function of the normalized input impedance (Zin) which can be expressed as shown below,

$$R = 20\log|\Gamma| = 20\log\left|\frac{Z_{in}-1}{Z_{in}+1}\right|$$

It can be seen from the above relation for Γ that the principle of the metal-backed single-layer absorber is to make use of the reflection reduction by impedance matching, in which the normalized input impedance with respect to the impedance of free space (Zin/Zo) should be 1 for no reflection.

Measurements of Permittivity, Permeability and Reflection Loss

Permittivity and permeability measurements can be carried out using network analyzer. A rectangular cavity, designed to obtain X-band frequencies (8.-12GHz) can be connected to the two ports of the network analyzer. S-parameter test set of the measuring system. It is operated in the TE_{10p} mode. The network analyzer generates the microwave oscillations of specific frequencies. With the help of an instrumentation computer, the measurements at different resonant frequencies can be carried out. For measuring permittivity values the following procedure can be employed.

The resonance frequency f_t and the unloaded quality factor Q_t of the cavity resonator can be measured with the empty capillary tube inserted in the cavity at the position of maximum electric field as shown in Figure 1. The sample fluid needs to be filled in the capillary tube and then sealed. It is positioned at the maximum electric field. The resonance frequencies f_s and loaded quality factor Q_s are measured. The inner diameter of the capillary tube is measured. and of the liquid sample can be calculated.

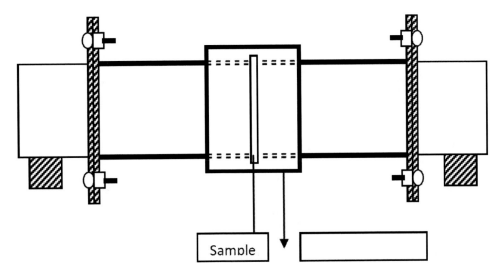

Figure 1. Cavity for the permittivity measurements (cylindrical cavity).

The procedure can be repeated for all the available resonant frequencies in the X band resonator.

Permeability measurements can also be performed using the same X band resonator. In this case, the sample material should be placed at the position of maximum magnetic field as shown in Figure 2.

Permeability and permittivity can be calculated using the equations

$$\varepsilon'_r + 1 = \frac{(f_t - f_s)V_c}{2f_s \; V_s} \quad (1)$$

$$\varepsilon''_r = \frac{V_c}{4V_s}\left(\frac{1}{Q_s} - \frac{1}{Q_t}\right) \quad (2)$$

Here, represent the quality factor of empty tube.

Similarly, the real and part of the complex permeability is given by,

$$\mu'_r + 1 = \left(\frac{\lambda_g^2 + 4a^2}{8a^2}\right)\frac{(f_t - f_s)V_c}{f_s \; V_s} \quad (3)$$

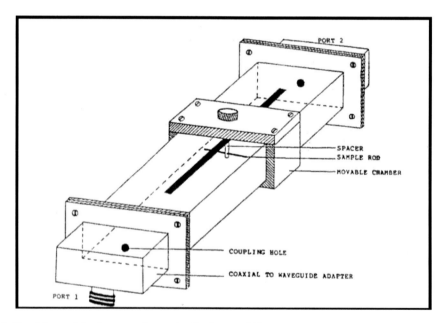

Figure 2. Top View of the Rectangular cavity for Permeability measurements.

Reflection loss can be calculated by employing the formula

$$Z_{in} = \left(\frac{\mu}{\varepsilon}\right)^{1/2} \tanh\left[j\left(\frac{2\pi f d}{c}\right)\right](\mu\varepsilon)^{1/2} \quad (4)$$

$$RL(dB) = 20\log\left|\frac{(Z_{in}-1)}{(Z_{in}+1)}\right| \tag{5}$$

where Zin is the input impedance, f is the frequency, d is the thickness and c is the velocity of light. Details are cited elsewhere [9].

Feng et al. [10] have conducted an investigation on the microwave absorption properties of barium ferrite/EPD rubber mixture and reported that real part and imaginary part of the complex permittivity for the barium ferrite/EPDMrubber absorbing materials are nearly independent of the frequency [10]. The permeability decreases initially with the increase of the frequency till a frequency of 4GHz is reached. Both real and imaginary part of the permeability increases from there to 8.2 GHz.

The real part of permeability has the maximum value at 8.2 GHz. Beyond 8.2 GHz, The μ'_r decreases as the frequency increases. The μ''_r spectrum is similar to the spectrum of μ'_r, and the two inflection points are 4.3 and 9.0 GHz. The natural resonance of the barium ferrite may lead to the frequency spectra of the μ'_r and μ''_r

Pant et al. [11] have studied the microwave absorption of ferrofluid-PVA-PANI (polyaniline) composite film and showed that the relative microwave absorption of such a film is greately influenced by the application of an external magnetic field. A relative absorption of 0.22 was obtained in presence of an applied magnetic field when compared to the absorption with out the magnetic field. This magnetic field assisted microwave absorption is explained as due to fast spin relaxation of super-paramagnetic particles distributed in polymer matrix. Kong et al. [12] studied the microwave absorption properties of magnetite, which is the most widely used magnetic constituent in ferrofluids.

In his study, he investigated the microwave propeties in a composite containing magnetite and thermoplastic natural rubber. As the ferrite filler increases in the composite of magnetite-thermoplastic natural rubber, the reflection loss also increases from -10.79dB (corresponds to 4wt% of the filler) to -25.51dB (corresponds to 12wt% of the filler). The absorbing band width is also found to be proportional the filler content. Band width increased to 2.7GHz for the sample with ferrite filler content of 12% compared to the 0.6% for that with 4%. Pure magnetite particles showed a minimum reflection loss of -32dB at 3.65 GHz and another reflection peak of -10.77dB at 11.65 GHz.

Swapna Nair et al. [13] studied that microwave dielectric properties and microwave reflection loss properties of Zinc, nickel and manganese substituted ferrfluids.

The microwave characteristics of the ferrofluid samples under investigation is shown in figure 5a to figure 5d.The samples show microwave absorption properties. Calculation of reflection loss based on eqns. (4) and (5) showed that the minimum reflection loss is -48 dB for the nickel substituted ferrofluids while a minimum reflection loss of -43dB is recorded for the zinc substituted ferrofluids.

Both ferrofluids had particle concentration levels of 0.30. A reflection loss less than - 25dB was recorded with in 1.5 GHz in the X band. The observed microwave absorption properties make them ideal candidates for microwave absorbers. The charaterisation in other microwave bands will provide information regarding the application potential of these fluids in microwave absorbing devices. Further studies are required in this direction to realize a ferrofluid based microwave absorber.

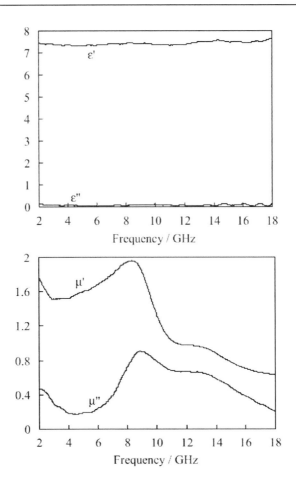

Figure 3. Permittivity and permeability of Barium ferrite (*Keng* et al.).

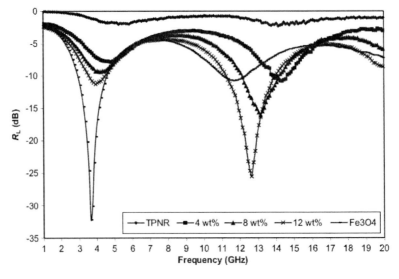

Figure 4. Frequency dependence of RL for pure TNPR, pure magnetite and their nanocomposites at the thickness of 9mm.

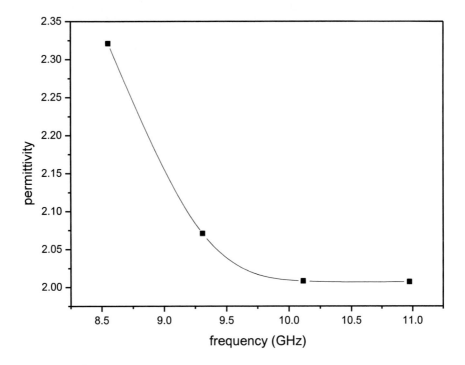

Figure 5a. Permittivity of $Zn_{0.4}Fe_{0.6}Fe_2O_4$ ferrofluid.

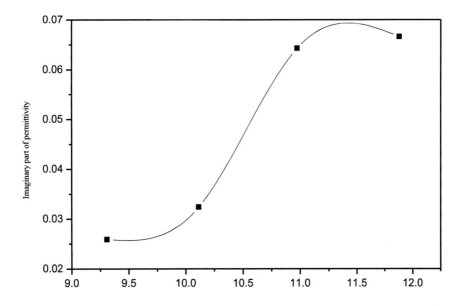

Figure 5b. Imaginary part of permittivity for $Zn_{0.1}Fe_{0.9}Fe_2O_4$ ferrofluid.

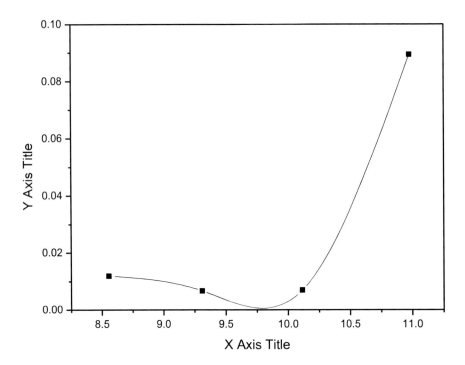

Figure 5c. Imaginary part of Permittivity for $Zn_{0.4}Fe_{0.6}Fe_2O_4$ ferrofluid.

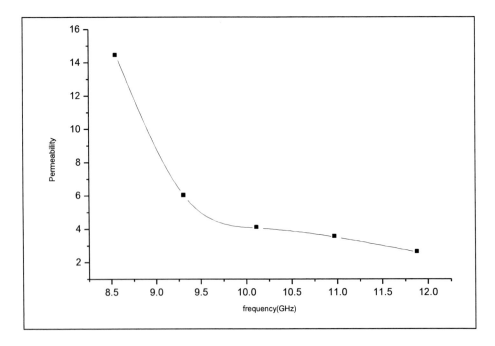

Figure 5d. Permeability for $Zn_{0.4}Fe_{0.6}Fe_2O_4$ Ferrofluid.

CONCLUSION

In conclusion this chapter describes the frequency, concentration, composition and magnetic field dependent electrical characteristics of various ferrites based ferrofluids.The increased demand for the microwave absorbing materials are the major motivation behind this study. The dielectric permittivity is found to be highly dependent on the dopent and concentration. The presence of an applied magnetic field found to change the dipolar polarization of this ferrofluids.The microwave characteristics of these systems were also estimated and the observed properties were highly encouraging.

REFERENCES

[1]   1 B. A. Larsen, M. A. Haag, M.H. B. Stowell, D.C. Walther, A. P. Pisano,C. R. Stoldt, *Proc. of SPIE*, 6525 (2007) 652519.

[2]   R. E. Rosenweig, *Ferro hydrodynamics*, Cambridge University Press, 1985.

[3]   B.M Berkovsky, V.S Medvedev., M.S Krakov, *Magnetic fluids; Engineering applications*, Oxford university press. 1993.

[4]   J. Deperiot, G. J Da Silva C.R Alves, *Braz. J. Phys.* 31, (2001), 390.

[5]   H. W Davies P. Llewellian, *J. Phys. D; Appl. Phys*, 13 (1980) 2327.

[6]   R.K Bhat, *Indian J. Eng. Mater. Scie.* 5, (Dec 1998), 477.

[7]   H.E Horng, C. Y. Hong, H.C. Yang, I J Jang, S.Y Yang, J.M Wu, S. l Lee, F. C Kuo, *J. Magn. Magn. Mater*, 201(1999) 215.

[8]   H. A. Bethe and J. Schwinger, NDRC, Rep. D1-117, 1943.

[9]   Y. Naito, K. Suetake, *IEEE Trans. Micro. Theory. Techn.* 19 (1971), 65.

[10] Y.B. Feng, T. Qiu, C.Y. Shen, *J. Magn. Magn. Mater* 318 (2007) 8.

[11] R.P. Pant, R. Rashmi, R.M. Krishna, P.S. Negi, K.Ravat, U. Dhawan , S.K. Dhawan, D.K. Suri , *J. Magn. Magn. Mater* 149, (1995) 10.

[12] Kong, S.H. Ahmad, M. H. Abdullah, D.H. Ahmad, N.Yusoff, D. Puryanti, *J. Magn. Magn. Mater* 322 (2010) 3401.

[13] S. S. Nair, M.R. Anantharaman, Microwave dielectric and absorbing properties of zinc substituted ferrofluids studied in the X band, *Mater.Chem.and Phys,* Submitted.

[14] S. S. Nair, F. Xavier M.R. Anantharaman, Microwave dielectric properties of ferrofluids based on $Ni_xFe_{1-x}Fe_2O_4$, *Mater. Lett,* Submitted.

In: Ferrofluids
Editors: Franco F. Orsucci and Nicoletta Sala

ISBN: 978-1-62808-410-8
© 2013 Nova Science Publishers, Inc.

Chapter 9

FERROFLUIDS BASED 1D MAGNETIC NANOSTRUCTURES

T. N. Narayanan[*]
Department of Mechanical Engineering and Materials Science
Rice University, Houston, US

ABSTRACT

Structured magnetic nanomaterials such as one dimensional (1D) magnetic nanostructures are highly influenced the fundamental and applied research due to their augmented magnetic activities and structural uniqueness. Ferrofluids contain magnetic nanoparticles in a carrier fluid, can be assembled inside or outside of other nanostructures to get a variety of structured nanomaterials having multiple functionalities. Decorating or infilling nanotubes and nanowires with magnetic nanoparticles caught the attention of scientists due to their extensive applications in various fields. Interest has been spread to the synthesis of various 1D nanostructures of both metals and nonmetals and hybridizing them with magnetic nanoparticles for various sensor and energy harvesting devices. The availability different functional groups in the organic hull surrounding the magnetic nanoparticles in ferrofluids make them easy to attach to the surface of nanotubes or nanowires. This chapter discusses various hybrid 1D magnetic nanostructures based on ferrofluids, their synthesis and applications. Since carbon is the most common, unique and stable coating material for magnetic nanoparticles, a special emphasis is given for carbon and ferrofluid based 1D magnetic nanostructures. A detailed account has been provided for the possible applications of such hybrid 1D magnetic nanostructures in biomedical and engineering fields.

1. INTRODUCTION

The realm of materials science underwent a sea change during the late 90's with the advent of nanoscience and nanotechnology. The pioneering efforts have recently revealed

---

[*] E mail: nt3@rice.edu, tn_narayanan@yahoo.com.

new physical and chemical insights of materials at a level intermediate between atomic/molecular and bulk, which are expected to make significant technological revolutions in the near future. The emergence and growth of nanostructured materials has actually been at an unprecedented rate, principally due to their unique and outstanding physical and chemical properties which could be controlled remarkably by tuning their morphology (i.e., size, shape, and dimensionality) [1-5]. A large number of prospective applications spanning wide areas such as optoelectronics, data storage, sensors, catalysis, energy storage, nanoelectronics, therapy, diagnosis and photochemistry could be realized due to the development of many innovative hybrid materials using these advances.

Recently, one dimensional (1D) nanostructures such as nanowires and nanotubes have also become the focus of intensive research owing to their unique applications in the fabrication of nanoscale devices [6-8]. 1D nanostructures got immediate attention soon after the landmark paper by Iijima [9] on carbon nanotubes in 1991 and various types of organic-inorganic 1D nanostructures were realised thereafter. 1D systems are the smallest dimensional structures that can be used for efficient transport of electrons/phonons and optical excitations, and are thus expected to be critical to the function and integration of nanoscale devices. However, little is known about the nature of carrier transport through a 1D system. Moreover, these systems should exhibit density of states singularities, for example Van Hove singularity in carbon nanotubes, can have energetically discrete molecule like states extending over large linear distances, which might show more exotic phenomena, such as the spin-charge separation predicted for a Luttinger liquid [10]. There are also many other applications where 1D nanostructures could be exploited, including nanoelectronics, superstrong and tough composites, functional nanostructured composites, and novel probe microscope tips [11-21].

Recent progress on magnetism and magnetic materials has made structured magnetic nanostructures as a particularly interesting class of materials for both scientific and technological explorations [22]. Studies on areas such as interlayer coupling, giant magnetoresistance, colossal magnetoresistance, tunneling magnetoresistance, exchange bias, half-metallic ferromagnets, spin-injection, and current induced switching have led to exciting possibility of utilizing electron spin for information processing and this paved a way to a new branch known as 'spintronics' [23, 24]. Novel properties will start to emerge as the sample size becomes comparable to or smaller that certain characteristic length scales such as spin diffusion length, carrier mean free path, magnetic domain wall width, superconducting coherence length etc. The effects of confinement, proximity, and degree of freedom govern the interplay between the relevant physical length scales and the sizes of the patterned magnetic materials [22]. Ordered magnetic nanostructures are particularly interesting to study, as one can probe both the individual and collective behaviour of the elements in a well-defined and reproducible fashion. Moreover, ordered magnetic nanostructures are important technologically because of their application potential in magnetic random access memory, patterned recording media, biomedical applications and magnetic switches [25, 26]. The assembly of atoms or other building blocks logically into structures with nanometer size and desired (low) dimensionality is a key challenge before material scientists. A better insight and understanding of major factors which control the growth of the nanostructures is indispensable for assembling these nano-building blocks. Ferrofluids are interesting nanomagnetic system having a range of applications from engineering to drug delivery [27-30]. Moreover, they are unique systems to study the fine particle magnetism and magnetic interactions among them. Various quantum confinement effects such as blue shift due to weak confinement and strain induced redshift [31] have been reported in ferrofluids and synthesis of various kinds of ferrofluids is still an active area of research. A lot of ferrohydrodynamaic modeling studies were carried out immediately after the first successful synthesis of ferrofluids by Rosenweig et al. [32]. The efforts to align/self-assemble ferrofluid nanoparticles to structured surfaces underwent during the last decade due to their fundamental and applied potential.

Self-assembling nanoparticles at surfaces and interfaces is an ingenious method to form structured materials from zero dimensional (0D) nanoparticles. Moreover, for realizing versatile functions, assembly of nanoparticles in regular patterns on surfaces and at interfaces is required [33]. The assembling of nanoparticles will generates new nanostructures those can have unforeseen collective and intrinsic physical properties and can be exploited for multipurpose applications such as in nanoelectronics, spintronics, various types of sensors etc. Self-assembly is an important tool for the fabrication of integrated nanoscale devices and it is the common feature used by the nature for the creation of multidimensional and complex structures such as cells [34]. Magnetically, optically or electrically nanoparticles can be assembled to form symmetrical structures having multiple functionalities. Magnetic nanoparticles can also be arranged in to 1D, 2D or 3D structures using a strong magnetic field, which will assemble the nanoparticles by strong inter-particle interactions [35, 36].

2. FERROFLUIDS

Ferrofluids are colloidal stable suspensions of ultrafine magnetic particles in a carrier fluid. The condition for arresting both agglomeration arising out of magnetic dipolar interaction and sedimentation due to gravity necessitates the size of magnetic particles to be of the order of few nanometers. The clustering of nanoparticles is further prevented by an organic hull surrounding each magnetic nanoparticles inside the suspension, which either stabilizes the particles via steric repulsion or electrostatic/columbic (ionic) repulsion. The first category fluids represent hydrocarbon based fluids while the latter one is water based fluids (aqueous). These organic hulls keep the ferrofluids stable, more than that, they help to make hybrid structures out of these ultrafine nanoparticles by covalent interactions. The ferrohydrodynamic conditions necessitate the size of nanoparticles so critical that the width of particle size distribution of suspended nanoparticles will be within 1-2 nm. The synthesis of monodisperse nanoparticles is one of the hurdles in experimental nanotechnology, and thus ferrofluids act as a perfect size selector.

This helps to form various magnetic nanocomposites from ferrofluids with uniform and repeatable magnetic properties, since the size distribution will affect the physical properties of the composite. Various nanofuids of different magnetic materials are realized and which includes different types of ferrite nanoparticles, metallic nanoparticles and hybrid nanomaterials. The characteristic sizes of most of the nanoparticles which constitute stable ferrofluids lie within 1 - 20 nm (figure 2).

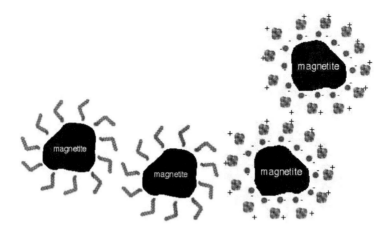

Figure 1. Schematic of stabilisation of magnetite nanoparticles in ferrofluid by two different mechanism; (left) steric repulsion (hydrocarbon based fluid), (b) columbic repulsion (aqueous fluid).

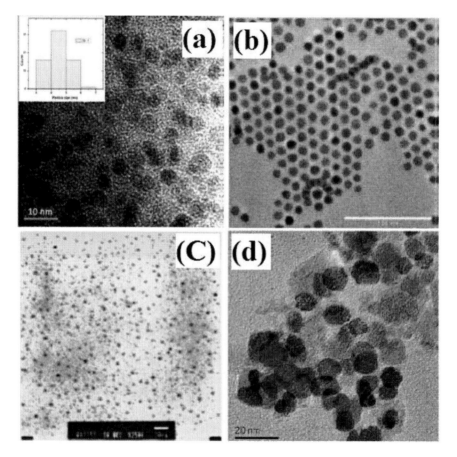

Figure 2. Transmission Electron Microscope images of various metallic and nonmetallic ferrofluid nanoparticles; (a) Ni, Ref:[28] (b) Co, Ref:[37] (c) Fe, Ref:[42] (d) Fe_3O_4 Ref:[29]. The nanoparticles are monodisperse and particle size is <20 nm.

3. FERROFLUIDS BASED SUPERSTRUCTURES

The synthesis of magnetic nanoparticle superstructures from ferrofluids demands the control over size and size distribution. Various wet chemical routes have been successfully demonstrated by various researchers in the recent past. Various fluids include that of monometallic magnetic nanoparticles of iron (Fe), cobalt (Co) and nickel (Ni) synthesized by electrochemical reduction, chemical reduction, thermal decomposition (conventional treatments and ultrasound), mechanical attrition and polyol process, fluids of multimetallic nanoparticles and nonmetallic ferrites [28, 37-41].

Various methods have been adopted for assembling magnetic colloids. All these methods depends the surface properties of nanoparticles and they differ by different preparation conditions.

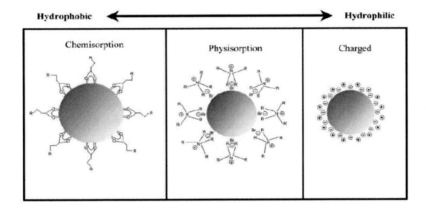

Figure 3. Sketches describing the particle surface at different conditions. The chemisorption, the dissociative adsorption of stabilizers, is most often used in hydrophobic solvents (ex: Toluene). The physisorption is used stabilize colloida by tetraalkyl ammonium salts in polar organic solvents (ex: THF). Charges surfaces are used in the stabilisation of aqueous suspensions. Ref: [37].

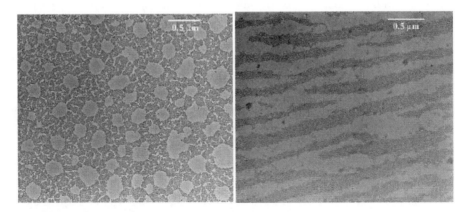

Figure 4. Self-assembled and field-assembled 2D structures; (left) TEM image of a sheet like self-assembled film obtained by drying a drop of ferrofluid on carbon coated Cu grid Ref: (right) TEM image of large area of chains deposited using ferrofluids by applying a magnetic field of 0.8T in perpendicular arrangement Ref: [44].

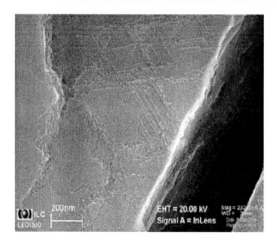

Figure 5. SEM image of a colloidal crystal prepared from a concentrated solution of highly monodisperse Co nanoparticles Ref: [37].

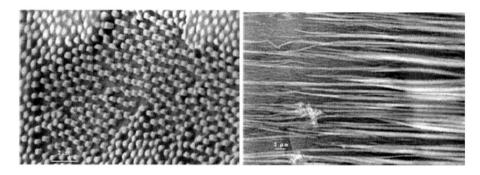

Figure 6. SEM images of, (left) randomly distributed rods prepared by drying the ferrofluid in air on substrate in a magnetic field of 0.8T perpendicular to the substrate, (right) rods by applying the field parallel to the substrate Ref: [45].

Various assembling methods include self-assembly, where self organization of particles into symmetric arrays take place during their deposition on flat surfaces, external field improved assembly, and layer-by-layer assembly [37]. Ferrofluid nanoparticles can be self-assembled in to 2D particle sheets having cubical or hexagonal ordered arrays depending on the stabilizer nature and/or the particle shape (figure 4).

The increase in concentration of ferrofluid will affect the particle area density and the number of layers formed. Again increase in concentration will result in to defected 3D crystals of various symmetries like hcp, fcc, or bcc. Such a 3D crystal synthesized by drying of a drop of monodisperse Co colloidal suspension is shown here [37].

Another interesting way to obtain supermolecular assemblies from ferrofluids is by magnetophoretic (magnetic field) deposition techniques. The size of these structures can be modulated by applied magnetic field. Nanorods grown either parallel or perpendicular to the substrate can be formulated by magnetic field assisted growth by applying parallel and perpendicular respectively. The increasing in concentration of colloids will result in to micron sized rod-shaped crystals (figure 6).

3.1. Ferrofluid Based 1D Nanostructures

Apart from the ferrofluid based micron sized rods by field assisted assembly, ferrofluids based nanostructures can be synthesized by assembling them over various nano-templates [46]. There are two types of templating methodologies, positive and negative templating, for the synthesis of ferrofluidic nanoparticles based 1-D nanostructures. Schematic of the two methodologies are shown in figure 7.

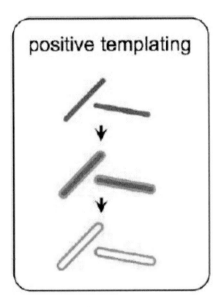
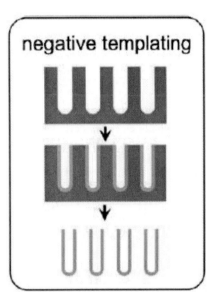

Figure 7. Two templating strategies; positive and negative templating. Template etching is mandatory in the case of negative templating Ref: [47].

In the case of positive templating, the functional nanoparticles are attached to the surface of various 1-D nanostructures such as nanowires, nanorods, or carbon nanotubes. Negative templating would rely on the objects having cylindrical pores; where the inside of those can be filled by nanoparticles.

Typical negative templates include porous anodic aluminum oxide (AAO), porous silicon, nanoporous gold thin films or track-etched polymers. Selective etching of the membranes will result in to bare functional nanoparticles based 1D structures.

3.1.2. Positive Templating

Biotemplates such as viruses/ protein strands are belong to the class of positive templates and are well suited for the large scale synthesis of structured nanomaterials from nanoparticles. A prime example of such a biotemplate is Tobacco mosaic virus (TMV)- a macromolecular complex consisting of a shell of 2130 coat protein monomers bound to a helical strand of RNA [46]. Solution based synthesis offers complete coating of the viruses with various materials whereas oxide nanoparticles can be directly precipitated on viruses. Metals can be precipitated on the surfaces via reduction of metal ions via electroless deposition.

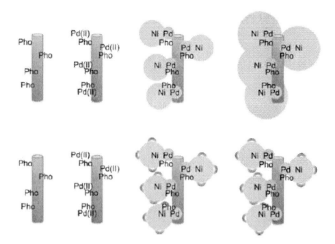

Figure 8. Schematic of electroless metallization of TMV. (top) Binding phosphate followed by binding Pd(II) complexes, reducing Pd(II) to Pd and growing Ni on Pd and continuous growth of Ni on Ni is demonstrate. (bottom) This sequence demonstrates the role of surfactant in inhibiting the growth. Growing Ni clusters are encapsulated in micelles consisting of amphiphilic surfactant molecules Ref: [47].

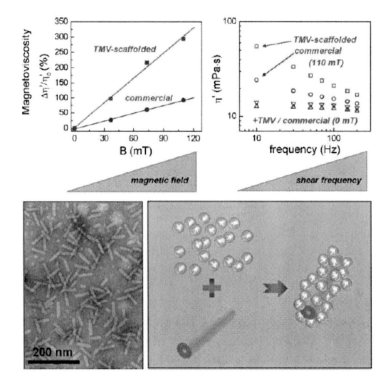

Figure 9. Demonstration of the enhancement in magneto viscosity of TMV-scaffolded ferrofluid (top left); improvement in the resistance to shear thinning (top right); SEM of genetically engineered nanotubes produced by the self-assebly of TMV coat protein about RNA strands of various lengths is shown in figure bottom left; Schematic of scaffolding of ferrofluid nanoparticles with virul protein surfaces is shown in figure bottom right Ref: [47].

The growth of nanoclusters can be hindered by addition of surfactants at proper time, thus we can form either thin layers or decoration on the biotemplates. The following figure shows an electroless metallization of TMV. Presence of noble metal clusters or particles is required for electroless deposition where they act as "sensitization" centers.

Figure 8 illustrates the typical morphological evolution of Ni on TMV during electroless deposition process. This also supports the theoretical prediction of mechanism of electroless deposition via "sensitization". Surfactants like Igepal in to the metallisation bath will result in to the complete suppression of cluster coalescence.

The same method can be adopted for the virul scaffolding of conventional ferrofluids. The main engineering application of templating the nanoparticles in ferrofluid is the change in their viscoelastic/magneto-viscous properties. The magneto-viscous properties of TMV scaffolded ferrofluids are changed drastically. Biotemplate ferrofluids exhibited an enhanced magnetoviscosity along with increased resistance to shear thinning. These changes in the physical properties are the result of "scaffolding" in which viral protein surfaces are interacted with magnetic nanoparticles of ferrofluids. The understanding of this mechanism also allows tailoring the phenomenon and thereby structuring the surface amino acids of the TMV by gene-technological alterations.

Modification of physical properties of ferrofluids by the addition of biological variants seems to be stable for several months, hence ensuring the stability of the structures and it demonstrates that biologically engineered nanoparticles have advanced past the stage of being mere laboratory curiosities [47].

Another protein based source for the positive templating is Collagen, a well know helical protein strands in animal skin. A team of researchers initiated by the author studied the synthesis of 1D nanostructures based on collagen and Iron oxide. Iron oxide nanoparticles found to be strongly bonded to the helical collagen fibers. The organic hull of citric acid in iron oxide is supposed to help in connecting the ultrafine iron oxide nanoparticles to collagen strands (unpublished data).

Moreover, collagen strands are thermally stabilized by these molecular interactions and the stabilisation of collage fibers is one of the prerequisites for the use of them in leather industry. So nanopatterning not only provides an ideal positive template nanostructures for possible magneto-rheological and sensor applications, it also meets an interesting industry requirement. Further, it also leads to a huge industry waste management, where a big industry waste can be converted in to a useful protein based magnetic nanocomposite having multiple functionalities. Similar to the biotemplates, magnetic nanoparticles from ferrofluids can also be decorated on the surfaces of carbon nanotubes and other metallic nanowires/nanotubes. Gold nanostructures are highly interested due to their bio-friendly, ease of functionalization nature and interesting photonic properties like plasmon emission. But they lack magnetic properties. Magnetically addressable gold nanorods will be an interesting candidate in nanotechnology. Such a goal has been accomplished by coating the gold nanorods with magnetite in an in-situ growth method [48]. Cetyl trimethylammonium bromide (CTAB) coated Gold nanorods [49] are again treated with single layer of polymer polystyrene sulphonate. Polystyrene sulphonate is an anionic material so it can be electrostatically bound with the cationic CTAB [50]. These coated nanorods are treated with a solution containing Fe(II) and F(III) ions in the molar ratio of 1:2 and the same procedure for Fe_3O_4 based ferrofluid is followed [29]. This will result in to magnetite coated Gold nanorods, which are magnetically active. TEM and magnetization curves are shown in figure 10.

CNTs have many potential applications; however technology development is in part hampered by the lack of effective means for manipulating these materials at very fine scales. Magnetic functionalization of CNTs will offer a potentially simple means for exerting this necessary control [51].

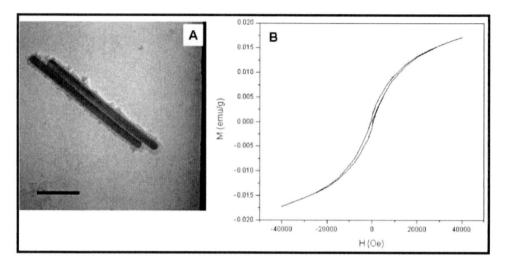

Figure 10. (a) TEM image of gold nanorods coated with a layer of iron oxide (scale bar=100 nm), (b) M(H) curve for iron oxide coated gold nanorods at T=2K Ref: [48].

There are many strategies developed for the hybridization of nanotubes with magnetic nanoparticles; some of them are particularly interested due to the fact that they do not destroy the innate surface chemistry of either particle. Surface functionalizations of tubes rely on the fact that post synthesized nanotubes can be functionalized with various groups by various wet chemical treatment. Since it is a post synthesis process, it can scale up too. A surface-preserving encapsulation of CNTs is feasible through even noncovalent bonding between nanoparticles and tubes.

In the absence of a chemical reaction or electrostatic bonding with surface defects, such a noncovalent bonding is happened through the minimization of the van der Waals potential between the spherical nanoparticles and cylindrical nanotubes. It is proven that for certain geometric and material parameter values; the van der Waals attraction is greatest between a sphere and cylinder than for either species to a neighbor of its own kind. In the case of defective CNTs (defects originated from harsh acid treatment/synthesis procedure) electrostatic interactions bind the nanoparticles with the surface of nanotube, but in this case the decoration may not be uniform.

TEM of non-aqueous ferrofluids (here magnetite) completely covering the CNTs is shown in figure 11. Here the nanoparticles are interacted with the nanotubes via noncovalent interactions.

The surfactants can also bring covalent interaction with nanoparticles and nanotubes and forming completely covered nanotubes with magnetic nanoparticles. Moreover, composite products obtained via these interactions can also stabilized by surfactants like CTAB or sodium dodecyl sulfate. Such a surfactant treatment will help to wrap several of these nanotubes to form micron sized cylinders. SEM image of such a bamboo shaped cylinder is shown in figur11 (right).

As it is mentioned before, some of the engineering fields also demand the synthesis of hybrid fluids with magnetic particles and CNTs. The ferrofluid-CNT composites in silicon oil found to exhibiting interesting magnetorheological properties and field dependent effects (figure 12) [51]. The strain modulation in the field on region is very clear from the graph, while in the field off region samples show consistent viscosity.

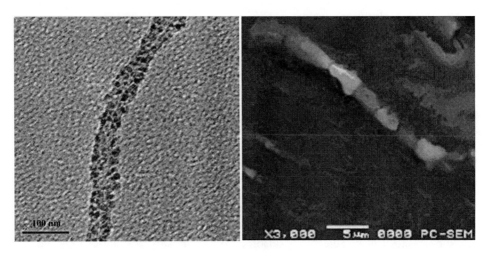

Figure 11. (left) TEM of organic ferrofluid nanoparticles bound to a CNT, (b) SEM of bamboo structured ferrofluid nanoparticles-CNT composite Ref: [51].

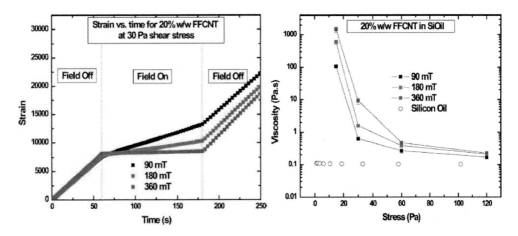

Figure 12. (left) Strain versus time plot for CNT-magnetite composite in silicon oil (composite nanofluid) for various applied fields, (right) viscosity versus stress for the composite nanofluid at three induction fields Ref: [51].

Irrespective of the applied field, the decrease in stress will diverge the viscosity and fluid become increasingly solid-like (figure 12-right). These viscometric properties demonstrate a range of rheological behavior that can be achieved by this class of non-Newtonian fluids [51]. Moreover, this combination can also be utilized for electrically and thermally conductive nanofluids, thus combining magnetic-electric-thermal-rheological fluid properties in a single hybrid system.

Another application of such a hybridization of magnetic particles with nanotubes is for the alignment of CNTs. These interactions between spherical particles and nanotubes can facilitate the alignment of CNTs in micro-electronic devices.

Demonstration of attempt to use of ferrofluid to forcibly align CVD-grown CNTs using a mechanical "raking" of the magnetic nanoparticles across the CNTs is shown in figure 13.

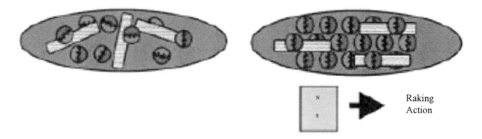

Figure 13. Alignment of CNTs using a ferrofluid Ref: [52].

3.1.3. Negative Templating

Template wetting and electrodeposition are the main techniques used in negative templating. Due to the high pore/surface energies inside the pores of nano channels like that AAO, nanoparticles can be selectively adsorbed on to the surfaces and etching the template will result in to 1D structures of adsorbed material. But template wetting based negative templating is hard to realise for the synthesis of stable structures from ferrofluids. But it can be achieved using some kind of other biotemplates such as magnetotactic bacteria. The advent on the controlled synthesis of nanocrystals opened the possibility of assembling them in custom made structures [53-55].

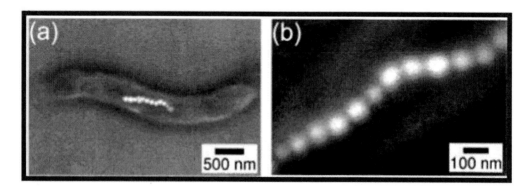

Figure 14. SEM images of magnetotactic bacteria, (a) bacteria synthesize a chain of magnetite nanoparticles, (b) close up of the chain of magnetite nanoparticles Ref: [56].

The manipulation of magnetic nanoparticles is also important for applications in spintronics, magnetic memory, and biology. It is already substantiated that either permanent magnets or electromagnets can be used to position and align superparamagnetic nanoparticles or magnetic nanowires/tube in fluids. Controlled assembly of the magnetic nanoparticles can also achieved by manipulating magnetotactic bacteria in a fluid with microelectromagnets [56]. Magnetotactic bacteria can synthesize a chain of nanoparticles inside their bodies and multiple group of bacteria can be assembled in a fluid using lithographically patterned micromagnets. After positioning them in an external magnetic field, the cellular membranes can be removed by cell lysis, leaving required pattern of magnetic nanoparticles on the substrate. Figure 14 shows the scanning electron micrographs of a magnetotactic bacteria named Magnetospirillum magnetotacticum that grows a chain of intracellular magnetite nanoparticles.

These bacteria can be used inside microfluidic chambers allowing them to align with the external magnetic field. The microfluid system controls the fluid flow. The magnetotactic bacteria can also used for the synthesis of monodispersed nanoparticles in fluid and then can be magnetically separated to form supermolecular structures.

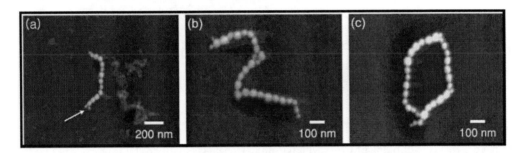

Figure 15. SEM images of assembled magnetic structures after removing cellular membrane of trapped bacteria, (a) single chain of magnetic nanoparticles is shown along with cellular debris, (b) a long chain of magnetite from a single bacterium. The chain bent during the cell lysis but remained intact due to the strong dipolar interactions. (c) A ring of nanoparticles formed by trapping and lysing two bacteria. The cellular bodies enclosing the magnetic chains prevent the clustering during the growth and cell lysis Ref: [56].

Yet another way to synthesize ferrofluids based 1D nanostructures is by nanocapillarity. Nanocapillarity is a complex phenomenon, the phenomenon of wetting the nanoporous channels by capillary force, and is a subject of intense investigation in leading laboratories all around the world [57]. It is a kind of negative templating method, but do not strictly follow the principle that template should be remove after the formation.

The filling of CNTs is accepted as an ingenious idea right from the early days of CNT research [58-60]. The confined existence of particles inside CNTs might introduce interesting new properties which are unseen in these systems previously [61]. Moreover filling of CNTs leads to confinement of particles and confinement of nanoparticles leads to possible quantum mechanical effects. Immediately after the theoretical prediction of capillary filling of CNTs by Pederson and Broughton [62], attempts were made to fill CNTs by capillary action of molten metals [63]

If the ferrofluids could fill the carbon nanotubes by nanocapillarity, it will be a versatile composite system to explore the confinement effects. The reports on biocompatibility of functionalised carbon nanotubes open another possibility of these composites in MRI contrast enhancement and magnetic field aided bio-medical drug delivery. The lack of complete understanding of liquid flow through nanochannels has propelled further activities on filling of CNTs by capillarity and their further characterizations [64, 65].

100% filling of MWCNTs using aqueous as well as non-aqueous iron oxide based ferrofluids can be achieved by capillary filling and manipulation of those magnetic nanotubes using external magnetic field. Figure 16 demonstrate the method of filling the aligned CNTs.

The so obtained CNTs are seems to be completely filled with iron oxide (once it is filled, carried fluid will evaporated off at room temperature) nanoparticles having size ~10 nm (figure 3). The penetration was found to be occurred almost simultaneously even without any external magnetic field and the applied external field only increase the rate of penetration.

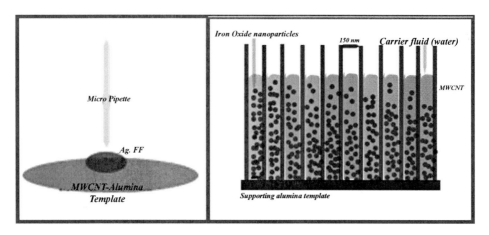

Figure 16. Schematic of filling of CNTs with ferrofluid by nanocapillarity Ref: [57].

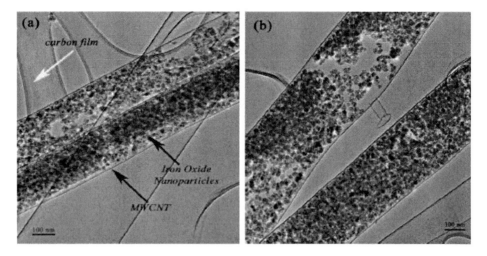

Figure 17. TEM images of Iron oxide filled MWCNTs Ref: [57].

The manipulation of magnetic nanotubes by external magnetic field has potential applications (figure 18).

These rotational effects of magnetic CNTs (figure18) even in low field prove that they can be used in MEMS applications and moreover the dipole-dipole interaction energy between them found to be higher than the energy of thermal excitations. Thus these magnetic CNTs are prone to forming chains as seen in figure 18b and 18c. The high magnetic response extends the applications of magnetic CNTs from wearable electronics to cantilever tips in magnetic force microscopes.

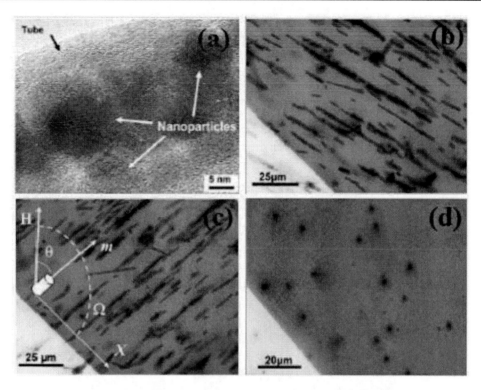

Figure 18. (a) High resolution TEM of Fe$_3$O$_4$ nanoparticles entrapped in MWCNTs, (b)-(d) manipulation of these magnetic nanotubes by magnetic field; (b) and (c) magnetic CNTs oriented along on the plane of supporting wafer, applied field=0.01T, (d) magnetic field applied perpendicular to the wafer, applied field=0.03T (nanotubes are frozen) Ref: [66].

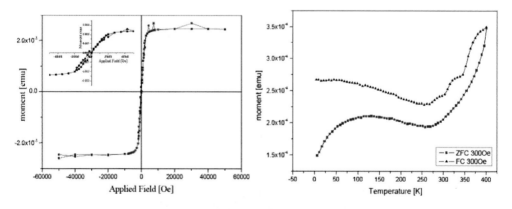

Figure 19. Magnetization studies on magnetite filled MWCNTs; (a) room temperature M(H) curve, (b) zero field cooling-field cooling curves. ZFC-FC curves showing anomalous magnetic properties due to the spatial confinement of nanoparticles inside MWCNTs Ref: [57].

Interesting magnetic properties are observed from CNT filled with magnetite composite [57]. The ZFC-FC curve shows an increasing magnetisation above room temperature showing its potential in field assisted applications and magnetic hyperthermia.

The most attractive applications of these 1D nanostructures are in magnetically aided drug delivery, bio-imaging and for diagnostics without surgical interference. Recent studies proved that magnetic carbon nanotubes with particle free surfaces have high drug loading capacity [67].

These CNTs are perfect template for field assisted diagnosis and drug targeting. Moreover, drug molecules can also be attached to the iron oxide surfaces. Selective sealing of the ends of CNTs is also possible using polymers, where it can be used for liquid delivery. CNTs are emerging as promising therapy-enhancing nanostructures due to their ability to move easily among the tissues and various compartments of the body and penetrate in to the cell, excellent stability in a biological environment, ultrahigh surface area with internal open space which will accommodate high amounts of therapeutic substances to be delivered in to the body [68-70].

Another advantage of graphitic systems is the ability to differentially modify the surface to avoid immune reactions, thus allow them to load with various substances for multimode actions such as delivery, targeting, and imaging. Various functional groups can be introduced in to the surface of nanotubes even employing harsh methods like acid treatment. Figure 20 shows the stability and morphology of the iron oxide after the nitric acid treatment. The concentrated nitric acid will allow the carboxylic group functionalization of CNT walls and this will enable the attachment of drug molecules to the surface. Functionalised CNTs (f-CNTs) are attracting the attention as new vectors for the delivery of therapeutic molecules.

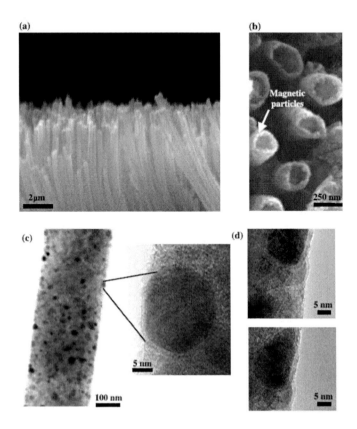

Figure 20. Various views of magnetic CNTs with particle free surfaces used for drug loading; (a) and (b) SEM images of cross section and top view respectively, (c) and (d) HR-TEM images of particle entrapped within graphitic wall Ref: [67].

As it is mentioned before, the ability of CNTs to cross cell membranes easily will help to deliver peptides, proteins and nucleic acids in to cells (both animal and plant cells). Different surface functionalization strategies also allow the selective drug attachment to surfaces [71].

Ferrofluid based graphitic materials can also address some of the environmental pollution related issues. Fe_3O_4-CNT composites can be used to remove heavy metal pollutants from water [72]. The adsorbed pollutants can be removed magnetically. The combination of graphitic nature and magnetic permeability allow magnetic CNTs for electromagnetic interference shielding and high energy radiation absorption [73, 74]. CNT based radiation preventing suites are available in the market. A simple method like nanocapillarity can introduce ferrofluids in to graphitic structures, and thus they can increase the shielding efficiency of the materials. The ferrite filled CNTs can be used to form magnetic Fe filled CNTs using high temperature reduction under hydrogen atmosphere [75], giving magnetic metal filled CNTs (X-ray diffraction pattern of such CNTs is shown in figure 21).

Ferrofluids not only make magnetic CNTs, but they can also act as catalysts for CNT growth. Nanocapillarity and template wetting properties of ferrofluids in 1D nanostructures such as nanotubes can be utilised to design and fabricate hybrid and highly porous nanostructures. They can find immense application from energy harvesting (supercapacitors and battery) to biomedical imaging and diagnosis. Immense research is going on the synthesis of ferrofluid based structured magnetic materials for various applications. Some of these fields include, plant science, biomedical imaging such as MRI (both T1 and T2 MRI imaging), targeted drug delivery, gene therapy, photonic and nonlinear optics, energy harvesting, water purification and magnetic data storage. Studies on the effect of magneto-elastic coupling of ferrofluid nanoparticles in various 1D matrices are also going on.

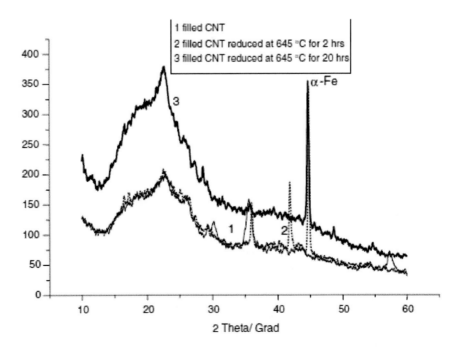

Figure 21. XRD of magnetite filled CNTs and αFe encapsulated CNTs after reduction of magnetite Ref: [75].

Thus size uniformity and surface functionalization of ferrofluids opened plethora of applications in various fields. Tuning the properties of ferrofluids can also achieved by various types of nanofillers available such as ferrites, mixed ferrites, monometallic, bimetallic and hybrid materials.

REFERENCES

[1] P. Harrison. *Quantum wells, wires and dots* (2000) John Wiely, New York.

[2] K. Barnham, D.Vvedensky, (Eds.) *Low dimensional semiconductor structures* (2001) Cambridge University Press, Cambridge.

[3] J. H. Davies, *The physics of low dimensional semiconductors* (1998) Cambridge University Press, Cambridge.

[4] C.Delerue, M.Lanno, *Nanostructures* (2004) Springer, New Delhi.

[5] G. Schmid, *Nanoparticles* (2004) Wiley-VCH, Weinheim.

[6] Z. L. Wang, *Adv. Mater.* 12 (2000) 1295.

[7] J. Hu, T. W.Odom, C. M. Lieber, *Acc. Chem. Res.* 32 (1999), 435.

[8] A Special issue in *MRS Bull.* 24 (1999), 20.

[9] S. Iijima, *Nature* 354 (1991) 56.

[10] C.Kane, L. Balents, M. P. A. Fisher, *Phys. Rev. Lett.* 79 (1997), 5086.

[11] M. Bockrath, D. H. Cobden, P. L. McEuen,; N. G. Chopra, A. Zettl, A. Thess, R. E. Smalley, *Science* 275 (1997) 1922.

[12] S. J. Tans, M. H. Devoret,; H. Dai,; A. Thess, R. E. Smalley, L. J. Geerligs, C. Dekker, *Nature* 386 (1997) 474.

[13] S. J.Tans, A. R. M.Verschueren, C. Dekker, *Nature* 393 (1998) 49.

[14] P.Yang, C. M. Lieber, *Science* 273 (1996) 1836.

[15] P.Yang, C. M. Lieber, *Appl. Phys. Lett.* 70 (1997) 3158.

[16] P.Yang, C. M.Lieber, *J. Mater. Res.* 12 (1997) 2981.

[17] H. Dai, J. H. Hafner, A. G. Rinzler, D. T. Colbert, R. E. Smalley, *Nature* 384 (1996) 147.

[18] S. S. Wong, J. D.Harper, P. T.Lansbury, C. M. Lieber, *J. Am. Chem. Soc.* 120 (1998) 603.

[19] P. M. Ajayan, J. M. Tour, Nature (2007) 447 1066.

[20] G. Schmid, *Clusters and Colloids: From theory to applications*, (1994) Wiley-VCH: New York.

[21] A Special issue in *MRS Bull.* 24 (1999), 20.

[22] J. J. Martin, J Nogues, K.Liu,; J. L. Vicent, I. K. Schuller, *J. Magn. Magn. Mater.* (Topical Review) 256 (2003) 449.

[23] G.Prinz, K. Hathaway, (Eds.) *Magnetoelectronics, Phys. Today* 48 (1995) (special issue).

[24] P. M. Levy, S Zhang, A. Fert, *Phys. Rev. Lett.* 65 (1990) 1643.

[25] R. P. Cowburn, *Science* 287 (2000) 1466.

[26] R. L. White , R. M. H. New, R. F. W Pease,. *IEEE Tran. Magn.* 33 (1997) 990.

[27] S. S Nair, T. Jinto, S. C. S. Sandeep, M. R. Anantharaman, P. Reji, *Appl. Phys. Lett.* 92 (2008) 171908.

[28] P. Reena Mary, S. C. S. Sandeep, T. N. Narayanan, P. Reji, M. Padraig, P. M. Ajayan, M. R. Anantharaman, *Nanotechnology,* 22 (2011) 375702.

[29] P. Reena Mary, T. N. Narayanan, S. Vijutha, D. Sakthi Kumar, Y. Yoshida, P. A. Joy, M. R. Anantharaman, *Nanoscale Res. Lett.* 5 (2010) 1706.

[30] T. N. Narayanan, A. P. Reena Mary, P. K. Anas Swalih, D. Sakthi Kumar, D. Makarov, M. Albrecht, P. Jayesh, A. Abdulaziz, M. R. Anantharaman, *J. Nanosci. Nanotech.* 11 (2011) 1958.

[31] S. S. Nair, M. Mathew, M. R. Anantharaman, *Chem. Phys. Lett.* 408 (2005) 398.

[32] R. E. Rosensweig, *Ferrohydrodynamics* Cambridge University Press, Cambridge (1985).

[33] S. Kinge, C. C. Mercedes, D. N. Reinhoudt, *Chemphyschem* 9 (2008) 20.

[34] G. Whitesides, B. Grzybowski, *Science* 295 (2002) 2418.

[35] J.J. Urban, D. V. Talapin, E. V. Shevchenko, C. B. Murray, *J. Am. Chem. Soc.* 128 (2006) 3248.

[36] M. Hilgendorff; B. Tesche, M. Giersig, *Aust. J. Chem.* 54 (2001) 407.

[37] M. Giersig, M. Hilgendorff, *Eur. J. Inorg. Chem.* (Microreview) 92005) 3571.

[38] B. Murray, S. Sun, H. Doyle, T. Betley, *Mater. Res. Soc. Bull.* 26 (2001) 985.

[39] M. Giersig, M. Hilgendorff, *J. Phys. D: Appl. Phys.* 32 (1999) L111.

[40] S. R. Hoon, M. Kilner, G. J. Russell, B. K.Tanner *J. magn. Magn. Mater.* 39 (1983) 107.

[41] N. Toshima, T. Yonezawa, *New J. Chem.* 22 (1998) 1179.

[42] K. T. Wu, Y. D. Yao, C. R. C. Wang, P. F. Chen, E. T. Yeh, *J. Appl. Phys.* 85 (1999) 5959.

[43] M. Hilgendorff, M. Giersig, in *nanoparticles assemblies and superstructures* (Ed: N Kotov) CRC press LLC, Boca Raton, (2005) 385.

[44] M. Giersig, M. Mulvaney, *J. Phys. Chem.* 97 (1993) 6334.

[45] Y. Chushkin, L. Chitu, S. Luby, E. Majkova, A. Satka, V. Holy, J. Ivan, M. Giersig, M. Hilgendorff, T. Metzger, O. Konovalov, *Mater. Res. Soc. Symp.* 877 (2005) S6-18.

[46] J. L. Lewandowski, in: *Virus Taxonomy* (Ed) C. M. Fauquet , M. A. Mayo , J .Maniloff, U. Desselberger, L. A. Ball, Elsevier Academic Press, Amsterdam (2005) 1009.

[47] W. Zhenyu, Z. Robert, M. Anna, S. E. Ruff, C. Ma, A. A. Khan, F. Geiger, B. A. Sommer, M. Knez, K. Nielsch, A. N. Bittner, C. Wege, C. E. Krill, *Phys. Status Solidi B* 10 (2010) 2412.

[48] J. Murphy, A. M. Gole, S .E. Hunyadi, C. J. Orendorff, *Inorg. Chem.* 45 (2006) 7544.

[49] N. R. Jana, L. Gerheart, C. J. Murphy, *Adv. Mater.* 13 (2001) 1389.

[50] Gole, C. J. Murphy, *Chem. Mater.* 17 (2005) 1325.

[51] S. Stephan, M. Gareth, *J.Fluid. Engg.* 129 (2007) 429.

[52] C. Smith, *"Ferrofluidic alignment of carbon nanotubes"*, NNIN REU Research Accomplishments (2006) 84.

[53] V. F. Puntes, K. M. Krishnan, A. P. Alivisatos, *Science* 291 (2001) 2115.

[54] S. Malynych, H. Robuck, G. Chumanov, *Nano Lett.* 1 (2001) 647.

[55] C. Wu, C. D. Chen, S. M. Shih, W. F. Su, *Appl. Phys. Lett.* 81 (2002) 4595.

[56] H. Lee, A. M. Purdon, V. Chu, R. M. Westervelt, *Nano Lett.* 4 (2004) 995.

[57] T. N. Narayanan, A. P. Reena mary, M. M. Shaijumon, C. Lijie P. M. Ajayan, M. R. Anantharaman, *Nanotechnology.* 20 (2009) 055607.

[58] B. M. Kim, S. Siha, H. H. Babu, *Nano Lett.* 4 (2004) 2203.

[59] R. E. Rosenweig, Ferrohydridynamics (1985) Cambridge University press 33.

[60] J. Popplewell, *Phys. Technol.*15(1984) 150.

[61] P. K. Tyagi, M. K. Singh, D. S. Misra, Filling of carbon nanotubes Encyclopedia of Nanoscience and *Nanotechnology*, 3 (2004) 417.

[62] M. R. Pederson, J. Q. Broughton, *Phys. Rev. Lett.* 69 (1992) 2689.

[63] G. Korneva,; H. Ye,; Y. Gogotsi, D. Halverson, G. Friedman J. C.Bradley, G.K. Konstantin, *Nano Lett.* 5 (2005) 879.

[64] T. Werder, J. H. Walther, R. L.Jaffe, T. Halicioglu, F.Noca, P. Koumoutsakos, *Nano Lett.* 1 (2001) 697.

[65] B. M. Kim, S.Sinha, H. H. Babu, *Nano Lett.* 4 (2004) 2203.

[66] G. Korneva, H. Ye, Y. Gogotsi, D. Halverson, G. Friedman, J.C .Bradley, K. G. Kprnev, *Nano Lett.* 5 (2005) 879.

[67] C. Vermisoglou, G. Pilatos, G. E. Romanos, E. Devlin, N. K. Kanellopoulos, G. N. Karanikolos, *Nanotechnology.* 22 (2011) 355602.

[68] K. Kostarelos, A. Bianco, M. Prato *Nature Nano.* 4 (2009) 627.

[69] A. Bhirde, V. Patel, J. Gavard, G. Zhang, A. A. Sousa, A. Masedunskas , R. D. Leapman, R. Weigert, J. S. Gutkind, J. F. Rusling, *ACS nano* 3 (2009) 307.

[70] Burke, X. F. Ding, R. Singh, R. A. Kraft, N. Levi-Polyachenko, M. N. Rylander, C . Szot, C. Buchanan, J. Whitney, J. Fisher, H. C. Hatcher, R. D. Agostino Jr., N D Kock, P. M. Ajayan, D. L. Carroll, S. Akmana, F. M. Torti, S. V. Torti, *Proced. Nat. Acad. Sci.* 106 (2009) 12897.

[71] W. Wei, W. Sebastien, P. Giorgia, B. Monica, K. Cedric, B. Jean-Paul, G. Renato, P. Maurizio, B. Alberto, Angew. *Chem. Int. Ed* 44 (2005) 6358.

[72] K.M. Asish, S. Ramaprabhu, *J. Phys. Chem.C* 114 (2010) 2583.

[73] T. N. Narayanan, V. Sunny, M. M. Shaijumon, P. M. Ajayan, M. R. Anantharaman, *ECS Lett.* 12 (2009) K21.

[74] K. S. Rajesh, T. N. narayanan , A. P. Reena Mary, M. R. Anantharaman, A. Srivastava, V. Robert, P. M. Ajayan, *Appl. Phys. Lett.* 99 (2011) 113116.

[75] D. Jain, R. Wilhelm, *Carbon* 45 (2007) 602.

In: Ferrofluids
Editors: Franco F. Orsucci and Nicoletta Sala

ISBN: 978-1-62808-410-8
© 2013 Nova Science Publishers, Inc.

Chapter 10

PROPOSED MULTIFERROIC/MAGNETO ELECTRIC FLUIDS: PROSPECTS, TECHNIQUES AND PROBABLE CANDIDATES

Swapna Nair[1], Radheshyam Rai[1] and K. Sudheendran[1,2]*
[1]Departmento de Engenharia Cerâmica e do Vidro and CICECO,
Universidade de Aveiro, Aveiro, Portugal
[2]Department of Physics, Sree Kerala Varma College, Thrissur, India

ABSTRACT

In this chapter, multiferroism is introduced and explained briefly. The former sections of the chapter dedicates to the different types of multiferroics, and the different ordering present in them. The role of symmetry in determining multiferroism is discussed. Different types of multiferroics are also introduced to the readers, which is grouped according to the type of ordering. Several examples for multiferroic materials in each group is provided. Although multiferroism means the simultaneaous possession of 2 or more ferroic order, the most sought after multiferroics are magneto-electric materials. Hence an introduction to the magneto electricity is also given in this section.

The latter section of the chapter, based on the discussions about possible multiferroic and magento-electric materials, proposes the synthesis of some multiferroic/magneto-electric fluids. These proposed candidates are superior smart fluids whose flow and properties can be both magnetically and electrically controlled. So they can be called electrically controllable ferrofluids. Hence the synthesis of such a fluid greatly enhances further, the application potential of ferrofluids.

INTRODUCTION TO MULTIFERROISM

Multiferroics can be primarily defined as single phase materials which simultaneously possess two or more primary ferroic viz. ferroelectric, ferromagnetic, ferroelastic and

---

* E-mail: swapna.s.nair@gmail.com.

ferrotoroidic properties. We owe greatly to H. Schmid for this particular definition [1]. Today the term multiferroic has been expanded to include materials which exhibit any type of long range magnetic ordering, spontaneous electric polarization, and/or ferroelasticity. Both single phasic as well as composite/multilayers that shows simultaneous muliferroic orders are now included in this definition. Extensive research is being carried out in unraveling different single phasic/composite materials who possess multiferroic ordering [2-19]. Considering the new expanded definition of multiferroism, the history of magnetoelectric multiferroics can be traced long back to the 1960s [20]

Hence, the field of multiferroics can be thought of originated primarily from the studies of magnetoelectric systems [21].

In 2003, large ferroelectric polarization was reported in epitaxial $BiFeO_3$ thin films [22] which showed a prominent jump towards the multiferroics for device applications. Strong magneto-electric coupling in orthorhombic $TbMnO_3$ [23] and $TbMn_2O_5$ [24] was reported following to the discovery of multiferroic $BiFeO_3$ thin films, which stimulated the recent research activities in the field of multiferroics.

SYMMETRY

Each multiferroic ordering is closely linked to symmetry. The primary ferroic properties can be characterized by their behavior under space and time inversion. Space inversion will reverse the direction of ferroelectric polarization P while leaving the magnetization M invariant. Time reversal, in turn, will change the sign of M, while the sign of P remains invariant.

Ferroelastic property is invariant in both space and time, while ferromagnetic property is only space invariant (time variant). Ferroelectric properties are time invariant, but space variant, while Ferrotorroidic properties are variant in both space and time.

Now consider, Magnetoelectric multiferroics. They require simultaneous violation of space and time inversion symmetry.

In $BiFeO_3$, the most studied single phasic multiferroic material, off-centering of ions breaks space symmetry giving rise to an electric polarization, while at a lower temperature additional magnetic ordering breaks time-reversal symmetry.

In general, a variety of mechanisms can cause lowering of symmetry resulting in multiferroicity as described below.

MULTIFERROICS - BY TYPE OF ORDERING

Charge Ordered

$LuFe_2O_4$, which shows improper ferroelectricity (i.e. no ionic displacement), below 330 K is one prominent example for a charge ordered multiferroic [25]. The charge ordering arises from the charge frustration on a triangular lattice with the mixed valence state of Fe^{2+} and Fe^{3+} ions. Ferrimagnetic behavior occurs below 240 K.

Generally, charge order can occur in a compound containing ions of mixed valence having geometrical /magnetic frustration. These ions form a polar arrangement, causing improper ferroelectricity (i.e. no ionic displacement). If magnetic ions are present, a coexisting magnetic order can be established and may be coupled to ferroelectricity.

Charge ordered ferroelectricity is suggested in Fe_3O_4 and $(Pr,Ca)MnO_3$ [26].

Magnetically Driven Ferroelectricity

Magnetically driven multiferroics are mostly oxides, in which macroscopic electric polarization is induced by magnetic long-range order. A necessary but not sufficient condition for the appearance of spontaneous electric polarization is the absence of inversion symmetry. In these materials inversion symmetry is broken by magnetic ordering. Such a symmetry breaking often occurs in so-called frustrated magnets, where competing interactions between spins favor unconventional magnetic orders. The microscopic mechanisms of magnetically induced ferroelectricity involve the polarization of electronic orbitals and relative displacement of ions in response to magnetic ordering.

Some multiferroics show the cycloidal spiral ordering, in which spins rotate around an axis perpendicular to the propagation vector of the spiral. The induced electric polarization is orthogonal to the propagation vector and lies in the spiral plane. The abrupt change of the spiral plane, induced by magnetic field, can result in a corresponding rotation of the polarization vector. Giant magnetocapacitance effect is reported in $DyMnO_3$ in which the transition is accompanied by 600% increase of dielectric constant [27]. The microscopic mechanism of magnetoelectric coupling in spiral multiferroics involves spin-orbit coupling [28].

Lone Pair Multiferroics

In usual perovskite-based ferroelectrics, the necessary criteria for ferroelecrticity is the "d0"ness. Here the transition metal ion (for example Ti in $BaTiO_3$, the mosinvestigated ferroelectric material) requires an empty "d" shell, since the ferroelectric displacement occurs due to the hopping of electrons between Ti "d" and O "p" atoms. The ferroelectric distortion occurs due to the displacement of B-site cation (Ti) with respect to the oxygen octahedral cage. This "d0"ness normally excludes any net magnetic moment because magnetism requires partially filled "d" shells. Partially filled "d" shell on the B-site reduces the tendency of perovskites to display ferroelectricity.

In order for the coexistence of magnetism and ferroelectricity (multiferroic), one possible mechanism is lone-pair driven [29] where the A-site drives the displacement and partially filled "d" shell on the B-site contributes to the magnetism. Examples include $BiFeO_3$, [30] $BiMnO_3$, [31] $PbVO_3$. In the above materials, the A-site cation ($Bi3+$, Pb^{2+}) has a stereochemically active $6s^2$ lone-pair which causes the Bi 6p (empty) orbital to come closer in energy to the O 2p orbitals. This leads to hybridization between the Bi 6p and O 2p orbitals and drives the off-centering of the cation towards the neighboring anion resulting in ferroelectricity.

Geometrically Frustrated Multiferroics

In this group of multiferroics, ordering is related to a structural phase transition at high temperature. $RMnO_3$ is a good example in this category. In hexagonal manganites like h-$RMnO_3$ (R=Ho-Lu, Y), the ferroelectric polarization at high temperature is correlated to lattice distortions through off-centering of ions. Geometric frustration gives rise to novel spin arrangements at low temperature giving ferromagnetic ordering: The spins order in a variety of non-collinear, e.g. (in-plane) triangular or Kagomé structures to relieve the geometric frustration. Hence the coexistence of ferroelectric and magnetic order occurs together with a strong coupling between them.

The mechanism of the ferroelectric ordering in hexagonal $RMnO_3$ is still under debate in the scientific community and must be understood before a comprehensive picture of multiferroic phenomena in spin frustrated systems can be built. It is still a discussion point whether the geometric distortion alone or the off-centering of Mn ions also contributes to the polarization. Physical properties of geometric multiferroics are dominated by the behavior of the d-shell electrons (eg-orbitals) and of the rare earth elements with an unfilled f-shell. Hexagonal manganites show the largest deviation from perovskite structure due to the small size of rare-earth ion. The strong coupling between ferroelectric and magnetic orders is represented by an anomaly in the static dielectric constant at magnetic phase transitions. Geometric frustrated ferroelectrics are prime candidates for device memory applications.

Several compounds belong to this important class of multiferroics: K_2SeO_4, Cs_2CdI_4, hexagonal.

MAGNETOELECTRIC EFFECT

The magnetoelectric (ME) effect is the phenomenon of inducing magnetic (electric) polarization by applying an external electric (magnetic) field. The effects can be linear or/and non-linear with respect to the external fields. In general, this effect depends on temperature. The effect can be expressed in the following form

$$Pi = \sum \alpha_{ij} H_j + \sum \beta_{ijk} H_j H_k +$$

$$Mi = = \sum \alpha_{ij} E_j + \sum \beta_{ijk} E_j E_k +$$

where P is the electric polarization, M the magnetization, E and H the electric and magnetic field, and α and β are the linear and nonlinear ME susceptibilities. The effect can be observed both in single phase and composite materials. One example of the important single phase magnetoelectrics is Cr_2O_3 [14]. Composite magnetoelectrics are combinations of magnetostrictive and electrostrictive materials, such as ferromagnetic and piezoelectric materials. The extent of the effect depends on the microscopic mechanism. In single phase magnetoelectrics the effect can be due to the coupling of magnetic and electric orders as observed in some multiferroics. In composite materials the effect originates from interface coupling effects, such as strain. Some of the promising applications of the ME effect are sensitive detection of magnetic fields, advanced logic devices and tunable microwave filters [32].

Applications

Multiferroic composite structures in bulk form are explored for high-sensitivity ac magnetic field sensors and electrically tunable microwave devices such as filters, oscillators and phase shifters (in which the ferri-, ferro- or antiferro-magnetic resonance is tuned electrically instead of magnetically [32].

In multiferroic thin films, the coupled magnetic and ferroelectric order parameters can be exploited for developing magnetoelectronic devices. These include novel spintronic devices such as tunnel magnetoresistance (TMR) sensors and spin valves with electric field tunable functions. A typical TMR device consists of two layers of ferromagnetic materials separated by a thin tunnel barrier ($\sim$2 nm) made of a multiferroic thin film [33]. In such a device, spin transport across the barrier can be electrically tuned. In another configuration, a multiferroic layer can be used as the exchange bias pinning layer. If the antiferromagnetic spin orientations in the multiferroic pinning layer can be electrically tuned, then magnetoresistance of the device can be controlled by the applied electric field [34]. One can also explore multiple state memory elements, where data are stored both in the electric and the magnetic polarizations.

Multiferroic/Magneto Electric Fluids

So far in this book, we have been discussing about ferrofluids. After discussing the simple theories and application of multiferroic materials, it will be very exciting to envisage a multiferroic/magnetoelectric fluid that combines the properties of a multiferroic material with added application potential of being a liquid. Again two fundamental types of them will come into our mind viz. single phasic and composite.

Single Phasic Multiferroic Fluids: Proposed Candidates

By this time, after reading this particular chapter, you know about different single phasic materials that possess multiferroic characteristics. Also the different synthesis schemes/methods to synthesize ferrofluids are already described in the former chapters. Then why not we envisage a synthesis technique to synthesize a single phasic multiferroic fluid?

First we have to think about the probable single phasic multiferroic candidates. $BiFeO_3$ and $BiMnO_3$ in which are lone pair multiferroic, are the most suitable candidates for being employed in the proposed single phasic multiferroic fluid. Co-precipitation is the most versatile technique employed for the synthesis of ferrofluids [35-39]. However, perovskites are little bit difficult to obtain through the cold chemicals precipitation techniques and the synthesised materials are often amorphous and hence the multiferroic ordering cannot be expected. Hence we will have to think about some other alternate synthesis techniques. Sol-gel technique is already employed for the synthesis of nanoscale ferrites and the formed particles are crystalline with phase purity [40-42]. Recently the synthesis of crystalline bismuth ferrite by sol-gel technique is reported [43, 44]. Hence sol-gel technique can be viable alternative for the synthesis of these multiferroic nanoparticles. However, insitu surfactant capping cannot be done using sol-gel technique. Hence sol-gel synthesised nanoparticles should be ball milled for low milling time with the desired surfactant and can be finally dispersed in the basefluid. Another promising candidate for these special smart fluids are $DyMnO_3$ in which a huge magneto capacitance is already reported. Charge ordered multiferroics like $LuFe_2O_4$ can also be employed. Quite interestingly, charge ordered multiferroism is predicted in magnetite (Fe_3O_4), which is widely employed for the synthesis of ferrofluids. However, not much effort have been taken to evaluate the multiferroic properties of magnetite based ferrofluids by researchers.

Magneto-Electric Composites

Single phasic mutiferroic materials exhibit only weak magneto electric coupling at room temperature and hence not ideal from the application point of view. Magneto-electric composites are viable alternatives for this. There are many possible magneto-electric pairs in which widely investigated candidates are PZT-NiF_2O_4 [46-49], and PZT-$CoFe_2O_4$ [50] pairs due to their excellent magneto electric properties at RT. But hazardous effects of lead based materials prompt us to seek after other magneto electric composites.

BZT- NiF_2O_4, BZT- CoF_2O_4, KNN- CoF_2O_4 etc are very good pair options. Zinc oxide – ferrite composites are another potential pair candidate for magneto electric fluid. However, the insitu synthesis of this particular composite is highly challenging due to the possible formation of Zinc ferrite. Magneto electric composites can be synthesised also by insitu sol-gel technique and further surfactation can be done by milling with the desired surfactant and finally these slurries can be dispersed in organic carriers /water. The advantage of employing these magneto-electric composites as the nanoparticle component in our proposed multiferroic fluid is the ease in their synthesis procedure. Single phasic multiferroism will be killed by some cation redistribution or some kind of polycrystallinity appears in our samples and hence the synthesis of a fluid with single phasic multiferroic material is a big experimental challenge. Additionally, their room temperature magneto-electric coupling is often very less. The above discussed composite systems can make up for both these disadvantage and we can hope that such a multiferroic fluid will be synthesised soon and investigated for their curious proper and electrically controllable fluid can be a reality which can have a vast application potential.

Synthesis of Multiferroic Fluids Based on Single Phasic Multiferroics

Nanosized bismuth ferrites and manganites can be synthesised by chemical methods such as co-precipitation and hydrothermal processes and sonochemical techniques [51-53]. Cr2O3 can also be synthesised by a hydrothermal process as described in [54]. The nanoparticles can be coated with the surfactant by ball milling with the desired surfactant for low milling time (~ 15 minutes) and the product can be dispersed directly in a carrier liquid.

Proposed Multiferroic/Magneto Electric Fluids

However the magneto electric properties especially the magneto electric coupling in these materials is very less at room temperature making their application potential lower as compared to their magneto electric composite counterparts.

Synthesis of Multiferroic Fluids Based on Magneto Electric Composites

Wet Milling

Magneto electric composites can be synthesised by a 3 step process.

- The wet milling of the piezo/ferroelectric material with the desired surfactant for 120 hrs
- Wet milling of the magnetic particle with the desired surfactant for 120 hrs
- Mixing of the two wet slurries and dispersing in any compatible carrier liquid

Disadvantage: This method can induce extensive local heat and surface defects which can affect the magneto electric coupling in the composite fluids.

Chemical Methods

A multiferroic composite consisting of nanoscale zinc oxide and magnetite can be individual co precipitation. Zinc oxide can be synthesised using the precursor materials zinc chloride, ethylene glycol and ethanol as described in [55]. Oleic acid or other surfactants can be added immediately upon the formation of nanoparticles. Surface coated magnetite can be separately synthesised by co-precipitation technique. Mixing of these two wet slurries in an organic/ water based carrier can be done after this step. Similar technique can be employed for the synthesis of BTO/ STO/BSTO – magnetite/NFO/CFO multiferroic fluids.

Sol gel technique can also be employed for this purpose.

CONCLUSION

The chapter envisages a fluid which can be controlled both electrically and magnetically and provides basic understanding of the types of magneto electric coupling and proposed candidates for realising such a fluid. Different tentative synthesis techniques are also provided. However, difficulties such as phase formation, grain growth, and improper coating of the surfactant can make the magnetic electric and magneto electric properties below our theoretical predictions. Scope exists for an elaborate research on the synthesis and properties of such fluids and the application potential are simply gigantic.

REFERENCES

[1]   H.Schmid, *Ferroelectrics* 162, 317 (1994).

[2]   G. Lawes and G. Srinivasan *J. Phys. D: Appl. Phys.* 44 (2011) 243001.

[3] G.Srinivasan, *Ann. Rev. Mater. Res.* 40, (2010) 153.

[4] Y. Tokura and S. Seki, *Adv. Mater.* 22, (2010)1554.

[5] K. F. Wang, J. M. Liu, and Z. F. Ren, *Adv. Phys.* 58, (2009) 321.

[6] Ce-Wen Nan, M. I. Bichurin, Shuxiang Dong, D. Viehland, and G. Srinivasan, *J. Appl. Phys.* 103, (2008) 031101.

[7] J. V. Brink and D. Khomskii, J. Phys.Cond. Matt. 20, (2008), 434217.

[8] T. Kimura, *Annu. Rev. Mater. Res.* 37, (2007), 387.

[9] R. Ramesh and N. A. Spaldin, *Nature Materials* 6, (2007), 21.

[10] C. N. R. Rao and C. R. Serrao, *J. Mat. Chem.* 17, (2007), 4931.

[11] S. W. Cheong and M. Mostovoy, *Nature Materials* 6, (2007)13.

[12] Y. Tokura, *J. Mag. Mag. Mat.* 310, (2007), 1145.

[13] M. Bibes and A. Barthelemy, *IEEE Trans. Electron. Dev.* 54, (2007), 1003.

[14] D. I. Khomskii, *J. Mag. Mag. Mat.* 306, (2006), 1.

[15] Y. Tokura, *Science* 312, (2006). 1481.

[16] W. Eerenstein, N. D. Mathur, J. F. Scott, *Nature* 442, (2006), 759.

[17] W. Prellier, M. P. Singh, P. Murugavel, *J. Phys.: Condens. Matter* 17, (2005), R803.

[18] M. Fiebig, *J. Phys. D: Appl. Phys.* 38, (2005), R123.

[19] N. A. Spaldin and M. Fiebig, *Science,* 309, (2005), 391.

[20] E. Ascher et al., *J.Appl. Phys.* 37, (1966), 1404.

[21] G. Smolenskii et al., *Sov.Phys. Solid State* 1 (1959)150.

[22] J. Wang et al., *Science* 299, (2003), 1719.

[23] T. Kimura et al., *Nature,* 426, (2003), 55.

[24] N. Hur et al., *Nature* 429, (2004), 392.

[25] N. Ikeda et al., *Nature* 436, (2005), 1136.

[26] S. W. Cheong and M. Mostovoy, *Nat. Mater.*, 6, (2007), 13.

[27] T. Goto et al., *Phys. Rev. Lett.* 92, (2004), 257201.

[28] M.E.Fisher and W.Selke, *Phys.Rev. Lett.* 44, (1980), 1502.

[29] N. A. Spaldin, *J. Phys. Chem. B* 104, (2000), 6694.

[30] J. B. Neaton, C. Ederer, U. V. Waghmare, N. A. Spaldin, K. M. Rabe, *Phys. Rev. B* 71, (2005), 014113.

[31] R. Seshadri, N. A. Hill, *Chem. Mater.* 13, (2001), 2892.

[32] C. W. Nan et al., *J. App. Phys.* 103, (2008), 031101.

[33] M. Gajek et al., *Nature Materials* 6, (2007), 296.

[34] C. Binek et al., *J. Phys. Cond. Mat.* 17, (2005), L39.

[35] L.F. López,. G.Bahamón, J. Prado, J.C. Caicedo, G. Zambrano, M.E. Gómez, J. Esteve, P. Prieto, *J. Magn.Magn.Mater,* 324, (2012), 394.

[36] S. S Nair, S Rajesh, V S Abraham and M R Anantharaman, *Bull.Mater.Scie,* 34, (2011), 245.

[37] S. S. Nair Francis Xavier. P.A. Joy, S.D. Kulkarni, M.R. Anantharaman, *J. Magn. Magn. Mater,* 320, (2008), 815.

[38] S. S. Nair, S. Rajesh, V.S. Abraham, M.R. Anantharaman, V.P.N. Nampoori, *J. Magn. Magn. Mater,* 305, (2006), 28.

[39] V.S. Abraham, S. S. Nair, S. Rajesh, U.S. Sajeev and M.R. Anantharaman, Bull. Mater. Scie, 27, (2004), 155.

[40] M. George, S.S. Nair, K.A. Malini, P.A. Joy, M.R. Anantharaman, *J. Phys. D: Appl. Phys.,* 40 (2007)1593-1602

[41] M. George, S.S. Nair, A.M. John, P.A. Joy, M.R. Anantharaman, *J. Magn.Magn.Mater.*302 (2006) 190.

[42] M. George, S.S. Nair, A.M. John, P.A. Joy, M.R. Anantharaman, *J. Phys. D: Appl. Phys.* 39 (2006) 900.

[43] T. Liu' Y. Xu' J. Zhao, *J. Amer. Cer. Soc*, 93, (2010) 3637.

[44] Z. Chen, G. Zhan, X. He, H. Yang, H. Wu, *Crys.Res.and Techn.*, 46, (2011)309.

[45] S. Lopatin, I. Opatina and I. Lisnevskaya *Ferroelectrics* 162 (1994), 63.

[46] J. Ryu, A. V. Carazo, K. Uchino H. E. Kim *J. Electroceram.* 7 (2001)17.

[47] G. Srinivasan, E. T. Rasmussen, J. Gallegos, R. Srinivasan, *Phys. Rev.* B 64 (2001), 214408.

[48] J. Y. Zhai, N. Chai and C. W. Nan *J. Phys. D: Appl Phys.* 37 (2004) 823.

[49] Q. H. Jiang, Z. J. Shen, J. P. Zhou, C. W. Nan *J. Eur. Ceram. Soc.* 27 (2007), 279.

[50] L. Weng, Y. Fu, S.Song, J. Tang, J.Li' *Scripta Materialia*, 56, (2007), 465.

[51] Gajović, S. Šturm, B.Jančar, A. Šantić, K. Žagar, M. Čeh,, *J. Amer. Cer. Soc.*, 93, (2010) 3173.

[52] T. P. Comyn, D.F. Kanguwe, J. He, A. P. Brown, *J.Eur.Cera.Soc.*28, (2008), 2233.

[53] N. Das, R. Majumdar, A. Sen, H. Maiti, *Mater.Lett.*, 61, (2007) 2100.

[54] Z. Pei, H. Xu, Y. Zhang, *J. Alloys and Comp.*, 468 (2009) L5.

[55] M. Moroni, D. Borrini, L. Calamai, L. Dei. *J. Coll.Inter. Sci .,* 286, (2005) 543.

INDEX

#

21st century, vii, 41

A

Abraham, 10, 36, 104, 146
access, 120
acetone, 8, 32
acid, 8, 33, 34, 79, 80, 81, 82, 83, 88, 91, 93, 127,
128, 134, 145
acidic, 32, 79
acrylonitrile, 34
active feedback, 65
actuation, 65, 108
adhesives, 99
adsorption, 123
adverse effects, 107
age, 35, 39
aggregation, 3, 108
alters, 18
aluminium, 14
aluminum oxide, 125
amino acid(s), 127
ammonia, 31, 34, 80
ammonium, 8, 31, 33, 81, 123
ammonium salts, 123
amplitude, 64, 65, 73
anisotropy, 17, 22, 23, 24, 26, 27, 32, 54, 87, 89, 90
antibody, 78
aqueous solutions, 68
aqueous suspension, 123
argon, 34
arrest, 100
Arrhenius law, 88
assessment, 56
atmosphere, 34, 98, 135

atoms, vii, 14, 15, 18, 19, 30, 40, 41, 43, 47, 103,
120, 141
attachment, 134, 135
Austria, 82

B

bacteria, 130, 131
bacterium, 131
band gap, 40, 41, 44, 45, 49, 70, 92
bandgap, vii, 48, 70, 92
bandwidth, 108
barium, 113
barriers, 26, 87, 88
base, 2, 3, 6, 7, 46, 54, 71, 73, 74, 86, 88, 100
beams, 56, 60, 63
benefits, 43
bias, 120, 143
biocompatibility, 77, 78, 79, 91, 131
biological samples, 78
biomedical applications, 29, 35, 79, 87, 88, 93, 97,
100, 120
biotechnology, vii
birefringence, 7, 54
bismuth, 144
bleaching, 45
Boltzmann constant, 87
Boltzmann distribution, 23
bonding, 41, 84, 128
bonds, 3, 84
bone marrow, 78
boundary surface, 109
breakdown, 69
Brownian motion, 4, 6, 54
building blocks, 120
bulk materials, 56

C

cancer, ix, 78, 79, 100
cancer therapy, ix, 78
cancerous cells, 79
candidates, ix, 15, 46, 56, 75, 78, 113, 139, 142, 143, 144, 145
capillary, 44, 109, 111, 131
carbon, 35, 69, 119, 120, 123, 125, 127, 131, 134, 137, 138
carbon nanotubes, 120, 125, 127, 131, 134, 137, 138
carcinoma, 92
catalysis, 27, 120
category a, 43
cation, 30, 31, 141, 144
cell death, 91
cell line, 92
cell membranes, 135
chain molecules, 2
chemical(s), viii, 1, 27, 29, 33, 40, 41, 42, 43, 45, 54, 68, 77, 78, 79, 80, 93, 99, 120, 123, 128, 143, 144
chemical properties, 41, 42, 78, 99, 120
chemical stability, 54
chemical vapour deposition, 43
chemisorption, 123
China, 11, 104
chromatography, 44
chromium, 15
circulation, 68
classes, 99
clustering, 121, 131
clusters, 7, 26, 74, 126, 127
coatings, viii, 64
cobalt, 2, 8, 14, 32, 34, 123
coherence, 61, 120
collage, 127
collagen, 127
collisions, 6
colloid particles, 2
color, 101
commercial, 11, 29, 34, 104
communication, ix, 65, 107
communication systems, ix, 107
community, 142
compatibility, 77, 91, 93
complexity, 65
composites, 56, 79, 120, 129, 131, 135, 144, 145
composition, viii, 2, 27, 41, 100, 108, 117
compounds, 2, 14, 54, 142
compression, 64
computer, 15, 111
computing, 60, 64
condensation, 43, 80, 81

conduction, 46, 67
conductivity, 100, 108
conductor, 8, 43
configuration, 58, 109, 143
confinement, viii, 42, 44, 45, 47, 48, 49, 120, 131, 133
conjugation, 78
consolidation, vii, 43
construction, 98
contamination, 31, 43
COOH, 34, 80
cooling, 87, 98, 99, 133
copper, 14
cost, 1, 15, 101
Coulomb energy, 42
Coulomb interaction, 47
covering, 128
crystal structure, 23, 29, 43
crystalline, 17, 32, 55, 144
crystallinity, 43
crystallisation, 35
crystallites, 43, 56
crystals, 18, 34, 47, 62, 80, 82, 124
culture, 82, 91
CVD, 43, 130
cycles, 64, 98

D

decomposition, 43, 68
decoration, 127, 128
defects, 31, 43, 100, 128, 145
deformation, vii, 109
degenerate, 56
degradation, 68
deposition, 43, 124, 127
depth, 41
derivatives, viii
detection, 78, 100, 143
deviation, 78, 142
dielectric constant, 7, 46, 63, 141, 142
dielectric permittivity, 62, 108, 117
diffraction, 57, 68, 75, 82, 84, 97
diffraction spectrum, 68
diffusion, 120
dimensionality, 120
dipole moments, 21
dipoles, 6, 21, 61
dispersion, 2, 7, 33, 43, 48, 80
displacement, 60, 63, 140, 141
dissociation, 8, 29
distilled water, 33, 80
distortions, 55, 58, 60, 142

distribution, 26, 31, 33, 34, 43, 44, 46, 47, 48, 77, 78, 79, 83, 86, 87, 88, 93, 101, 121, 123
DOI, 10, 36
doping, 49
dosage, 78
dream, 1, 2, 49
drug delivery, vii, ix, 9, 78, 83, 97, 100, 120, 131, 133, 135
drugs, 78
drying, 123, 124
dyes, 92
dynamic control, 65

E

electric charge, 68
electric field, 60, 61, 62, 63, 68, 111, 143
electrical properties, 108
electricity, 139
electrodeposition, 130
electroless deposition, 125, 127
electromagnetic, 60, 66, 107, 108, 109, 110, 135
electromagnetic fields, 108, 109
electromagnetic waves, 110
electron(s), 13, 14, 15, 19, 42, 44, 46, 47, 48, 49, 60, 66, 67, 81, 83, 120, 131, 141, 142
electron microscopy, 81
electronic structure, 6
emission, 40, 127
employment, 16, 41, 46, 98
emulsions, 100
encapsulation, 91, 128
energy, 3, 4, 5, 6, 7, 15, 16, 17, 19, 22, 23, 24, 25, 26, 29, 30, 33, 42, 43, 44, 45, 47, 48, 49, 54, 57, 60, 64, 66, 67, 69, 70, 71, 81, 83, 87, 88, 92, 99, 119, 120, 132, 135, 141
energy density, 66
energy input, 99
engineering, vii, 41, 54, 119, 120, 127, 129
environment(s), 26, 98, 107, 134
enzyme, 78, 82, 91
enzyme immobilization, 78
equilibrium, 23, 25, 60, 68
equipment, 8, 15, 98
etching, 125, 130
ethanol, 81, 145
ethyl alcohol, 80
ethylene, 145
ethylene glycol, 145
evaporation, 2
evidence, 17, 23, 87
evolution, 127
excitation, 44, 46, 53, 64, 67, 71, 72, 73, 74, 75, 92

exciton, 45, 46, 47, 48, 49
exclusion, 44
experimental condition, 74, 90
exposure, ix, 98, 108
external magnetic fields, vii
extinction, 7

F

fabrication, 41, 43, 100, 120, 121
ferrite, 30, 32, 113, 114, 121, 135, 144
ferrite composite, 144
ferroelectrics, 141, 142
ferromagnetic, ix, 2, 7, 14, 15, 16, 18, 19, 21, 22, 79, 100, 139, 140, 142, 143
ferromagnetism, 13, 14, 19
ferromagnets, 14, 15, 120
fibers, 127
films, 26, 100, 101, 140
filters, 97, 100, 101, 143
fine tuning, viii
fish, 32
flexibility, 56
flow control, viii
fluctuations, 6, 26, 60
fluid, viii, ix, 1, 2, 3, 5, 6, 7, 9, 32, 33, 34, 35, 46, 49, 54, 71, 73, 74, 77, 78, 79, 80, 81, 82, 84, 86, 90, 93, 98, 99, 100, 104, 111, 119, 121, 122, 129, 130, 131, 132, 139, 143, 144, 145
force, 5, 6, 79, 81, 99, 131
formation, 7, 8, 32, 34, 35, 41, 46, 81, 84, 107, 131, 144, 145
formula, 29, 68, 83, 112
four-wave mixing, 56
freedom, 42, 120
freezing, 26
friction, 5, 99
FTIR, 85
functionalization, 127, 128, 134, 135, 136

G

gel, 43, 143, 145
gene therapy, 135
genetic engineering, vii
geometry, 57, 58, 59, 67
glasses, 64
glycol, 77, 79
god schematic diagram, viii
google, 93, 102
grain size, vii, viii, 26, 31, 39, 40, 41, 42, 43, 44, 45, 46, 47, 48, 49, 50

graph, 129
graphite, 14
gratings, 97
gravitational field, 5
gravitational force, 6
gravity, 34, 80, 99, 121
growth, 7, 22, 43, 44, 87, 120, 124, 126, 127, 131, 135, 145

H

Hamiltonian, 47
harmful effects, 107
harvesting, 119, 135
heat capacity, 8
heat transfer, 32, 99
heavy particle, 6
hexane, 32
history, 140
homes, 102
hormone, 78
host, 49
House, 11, 104
housing, 98
human, vii, ix, 40, 54, 92, 107
human body, 107
human papilloma virus (HPV), 92
humidity, 99
hybrid, 119, 120, 121, 129, 135, 136
hybridization, 49, 128, 130, 141
hydrocarbons, 3, 99
hydrogen, 46, 47, 135
hydrolysis, 80, 81
hydrothermal process, 144
hydroxide, 8, 31, 33, 81
hyperthermia, vii, ix, 9, 77, 78, 79, 83, 89, 90, 93, 97, 100, 133
hysteresis, 15, 16, 20, 23, 79, 86, 87, 89
hysteresis loop, 15, 16, 20, 86, 87

I

ideal, viii, 7, 15, 41, 46, 54, 65, 101, 107, 113, 127, 144
identification, 94
image(s), 83, 84, 85, 93, 102, 122, 123, 124, 128, 130, 131, 132, 134
imaging systems, 100
immune reaction, 134
in vitro, 79
in vivo, 79
India, 1, 13, 27, 29, 39, 53, 77, 101, 107, 139

induction, 129
industry, viii, 42, 97, 127
information processing, 120
infrared spectroscopy, 81
initial state, 24, 66
insertion, 67, 109
integration, viii, 120
interface, 26, 71, 142
interference, 15, 77, 101, 133, 135
intraocular, 78
inversion, 55, 63, 140, 141
ions, 18, 26, 30, 32, 33, 127, 140, 141, 142
iris, 65
iron, 2, 8, 14, 15, 17, 33, 34, 77, 78, 79, 80, 81, 82, 83, 84, 85, 86, 87, 88, 90, 91, 92, 100, 123, 127, 128, 131, 132, 134
irradiation, 43
issues, 135

J

justification, 48

K

keratin, 92
kerosene, 2, 32, 33, 53, 71, 72, 73, 74, 75
kinase activity, 92
kinetics, 43

L

laser radiation, 54, 61
lasers, 61, 101
laws, 4, 41
lead, 30, 46, 67, 68, 93, 113, 144
leakage, 98, 100
lens, 56, 57, 67, 69
lifetime, 66, 69
light, 47, 55, 56, 60, 61, 62, 64, 65, 66, 67, 68, 69, 100, 101, 113
light scattering, 69
light transmission, 69, 101
liquids, 5, 8, 32, 55, 63, 78, 99
liver, 79
low temperatures, 15, 26, 32
lubricants, ix, 97, 99
luminescence, 92
Luo, 10, 75
lying, 39
lysis, 131

Index

M

magnet, 4, 9, 14, 16, 32, 33, 98, 99
magnetic characteristics, 1
magnetic field, viii, 2, 4, 5, 7, 8, 9, 13, 14, 15, 16, 17, 21, 22, 45, 54, 77, 78, 79, 81, 82, 88, 89, 93, 97, 98, 99, 100, 101, 108, 112, 113, 117, 121, 123, 124, 131, 132, 133, 141, 142, 143
magnetic field effect, 7
magnetic force microscope, 132
magnetic materials, vii, viii, 1, 14, 16, 27, 32, 35, 40, 41, 42, 45, 49, 120, 121, 135
magnetic moment, 14, 15, 19, 22, 23, 26, 30, 54, 78, 87, 89, 91, 99, 141
magnetic nanomaterials, viii, 119
magnetic particles, viii, 2, 3, 8, 30, 43, 46, 54, 78, 80, 81, 83, 87, 98, 121, 129, 130
magnetic properties, viii, 1, 2, 5, 13, 15, 26, 27, 29, 39, 40, 41, 42, 45, 49, 77, 80, 88, 104, 108, 121, 127, 133
magnetic relaxation, 87
magnetic resonance, 78, 79, 100, 143
magnetic resonance imaging, 78, 100
magnetic structure, 27, 131
magnetism, viii, 1, 13, 15, 27, 46, 49, 120, 141
magnetization, 5, 7, 14, 15, 16, 17, 18, 20, 21, 22, 23, 24, 25, 26, 27, 40, 42, 46, 54, 77, 79, 86, 87, 88, 90, 93, 99, 127, 140, 142
magneto-electric materials, ix, 139
magnetoresistance, 120, 143
magnets, 1, 15, 16, 98, 102, 130, 141
magnitude, 17, 19, 27, 55, 58, 61, 63, 64, 67
malignancy, 78
malignant cells, 79
manganese, 8, 14, 49, 113
manganites, 142, 144
manipulation, 40, 130, 131, 132, 133
mapping, 84, 85
mass, 5, 8, 16, 46, 47, 48, 49, 84
materials science, vii, 119
matrix, 26, 49, 92, 107, 108
matter, 6, 60, 107
measurement(s), 23, 27, 55, 59, 60, 69, 71, 75, 77, 81, 83, 86, 87, 88, 90, 92, 93, 111, 112
media, 2, 27, 55, 56, 61, 67, 68, 69, 120
medical, vii, 2, 78, 88, 131
medicine, viii, ix, 100
melting, 1, 41
membranes, 125, 131
memory, 15, 120, 130, 142, 143
MEMS, 100, 132
metabolism, 79
metal ion(s), 15, 125

metal nanoparticles, 34, 35, 43, 69, 73
metal salts, 32, 35
metals, vii, 31, 34, 35, 46, 49, 98, 119, 131
meter, 69
microemulsion, 32
microscope, 120
microwave radiation, 107
microwaves, ix
military, ix, 107, 108
miniature, 99, 100, 101
mitochondrial damage, 91
mixing, 30, 31, 33, 34, 43
models, 46
modifications, viii
modules, 65
moisture content, 81
molecules, vii, 3, 4, 6, 7, 14, 15, 41, 43, 54, 68, 78, 79, 108, 126, 134
momentum, 13, 30, 46, 47, 48
monolayer, 83
monomers, 125
morphology, 27, 108, 120, 134
mosaic, 125
motivation, 41, 117
MR, 79
MRI, 9, 78, 97, 131, 135
multidimensional, 121
multiplier, 69
museums, 102
music, 102

N

NaCl, 32
nanocomposites, 40, 114, 121
nanocrystals, 44, 45, 46, 130
nanoelectronics, 120, 121
nanomaterials, vii, viii, 44, 80, 108, 119, 121, 125
nanometer(s), vii, 42, 56, 120, 121
nanometer scale, vii
nanoparticles, 8, 32, 34, 35, 40, 44, 45, 49, 54, 77, 78, 79, 80, 81, 83, 84, 85, 87, 88, 91, 92, 93, 104, 119, 121, 122, 123, 124, 125, 126, 127, 128, 129, 130, 131, 132, 133, 135, 137, 144, 145
nanorods, 125, 127, 128
nanostructured materials, 43, 44, 120
nanostructures, v, vii, 119, 120, 121, 125, 127, 131, 133, 134, 135, 136
nanotechnology, vii, 39, 40, 41, 45, 50, 100, 119, 121, 127
nanotube, 69, 128
nanowires, 42, 119, 120, 125, 127, 130
National Aeronautics and Space Administration, 2

Nd, 69
NEMS, 100
neutral, 77, 79
next generation, 108
nickel, 2, 8, 14, 32, 34, 35, 49, 53, 113, 123
non linear absorptions, viii
nonlinear optics, 135
normal distribution, 83, 86
nucleating agent, 35
nucleation, 44, 88
nuclei, 19
nucleic acid, 135
nucleus, 60
null, 57

O

OH, 33, 34, 80, 81, 84
oil, 3, 32, 98, 129
oleic acid, 2, 3, 32, 53, 54, 75, 79
one dimension, 119, 120
optical communications, 64
optical limiters, viii, 54, 64, 65, 66, 68
optical parameters, 60
optical properties, vii, viii, 7, 39, 40, 42, 44, 45, 46, 47, 49, 53, 54, 55, 60, 91, 92
optical pulses, 65
optical systems, 64
optimization, 78
optoelectronics, 120
orbit, 13, 17, 141
organic compounds, 94
organic solvents, 123
organs, 107
oscillation, 68
oscillators, 143
overlap, 18, 19
oxidation, 31, 34
oxide nanoparticles, 77, 78, 79, 81, 83, 84, 86, 87, 88, 92, 100, 125, 127
oxygen, 14, 30, 84, 141

P

p53, 92
parallel, 7, 14, 15, 18, 19, 21, 25, 62, 84, 124
passivation, 34
pathogens, 78
peptides, 135
permeability, 5, 14, 15, 17, 108, 109, 110, 111, 112, 113, 114, 135

permittivity, 61, 107, 108, 109, 110, 111, 112, 113, 115
petroleum, 32, 33
pH, 32, 34, 43, 44, 77, 79, 80, 81
phagocytosis, 79
phase conjugation, 55
phase transitions, 142
Philippines, 94, 95
phonons, 120
phosphate(s), 44, 126
photolithography, 101
photoluminescence, 91, 92
photonics, vii
photons, 61, 67, 69
physical characteristics, 65
physical laws, 1, 4, 9
physical properties, 42, 43, 121, 127
physics, viii, 18, 40, 41, 55, 136
platform, 26, 27, 78
playing, 39, 49
polar, 3, 54, 123, 141
polarizability, 61, 62
polarization, 60, 61, 62, 63, 64, 117, 140, 141, 142
polio, 92
pollutants, 135
pollution, 135
polydispersity, 101
polymer(s), 34, 44, 113, 125, 127, 134
polymer matrix, 113
polymeric materials, 34
polymerization, 81
polystyrene, 127
population, 67
Portugal, 1, 13, 29, 39, 53, 97, 107, 139
precipitation, 1, 2, 29, 31, 32, 33, 35, 43, 53, 68, 75, 77, 79, 80, 93, 143, 144, 145
preparation, 11, 32, 43, 46, 54, 78, 104, 123
prevention, 5
principles, 43, 56
probability, 23
probe, 92, 120
propagation, 55, 62, 141
proportionality, 16, 17, 24
proposition, 50
protection, 54, 66, 68
protein kinase C, 92
proteins, 135
purity, 83, 144
PVA, 113

Q

quantization, 42, 44, 47

Index

quantum confinement, vii, viii, 42, 44, 45, 47, 48, 49, 120

quantum dot(s), 41, 42, 45, 46, 47, 48

quantum well, 42, 47

quartz, 86

quasiparticles, 46

R

radar, 108

radial distance, 69

radiation, 60, 64, 67, 68, 69, 92, 107, 135

radio, 77, 107

radius, vii, viii, 6, 41, 42, 44, 45, 46, 47, 48, 49, 57, 89

random media, viii

rare earth elements, 142

reactants, 8, 32, 44

reaction temperature, 44

reactions, 44

reactive sites, 41

reactivity, 41

reading, 9, 49, 60, 143

reality, 144

recalling, 26

redistribution, 31, 144

redshift, 121

refractive index, 55, 57, 58, 59, 60, 61, 62, 63, 64, 67, 68, 69, 73

refractive indices, 7

relaxation, 22, 23, 24, 78, 79, 89, 93, 113

relaxation process(s), 79, 89

relaxation rate, 23

relaxation times, 24

relevance, 91

repair, 78

repulsion, 3, 8, 79, 121, 122

requirements, 2

researchers, 1, 45, 91, 101, 103, 123, 127, 144

resistance, 75, 99, 126, 127

resolution, 133

resonator, 111, 112

response, 16, 55, 60, 61, 65, 66, 67, 77, 132, 141

Richard Feynman, vii, 39, 41

rings, 98

RNA, 125, 126

rods, 15, 124, 125

room temperature, 33, 80, 86, 87, 132, 133, 144, 145

rotations, 30

roughness, 41

routes, 123

rubber, 113

S

saturation, 16, 23, 27, 34, 46, 54, 55, 59, 77, 87, 93, 99

scattering, viii, 53, 66, 68, 69, 73, 74, 75

science, vii, 40, 102, 103, 135

scope, 40, 42

sediment(s), 34, 80

sedimentation, 3, 30, 79, 82, 121

self-assembly, 124

semiconductor(s), vii, 40, 42, 44, 45, vii, 40, 54, 56, 64, 67, 70, 73, 136

sensing, 65, 78, 100

sensitivity, 56, 58, 59, 143

sensitization, 127

sensors, 15, 54, 79, 97, 100, 108, 120, 121, 143

shape, 23, 26, 58, 64, 78, 120, 124

shear, 126, 127

shelf life, viii, 34, 75, 77, 82

showing, vii, 69, 83, 133

side effects, 78, 79

signals, 73

silanol groups, 79

silica, 77, 79, 80, 81, 83, 84, 85, 87, 88, 91, 92, 93, 109

silicon, 2, 125, 129

silver, 42

simulation(s), 81, 90

single chain, 131

sintering, 30

SiO_2, 81, 84, 85, 87, 88, 91, 92

SiO_2 surface, 88

skin, 127

smart materials, 100

sodium, 34, 35, 128

sol-gel, 35, 43, 80, 144

solid state, 30

solution, 8, 31, 32, 33, 34, 35, 48, 80, 82, 102, 124, 127

solvents, 123

species, 79, 128

specific gravity, 7

specific heat, 73, 79

spectroscopy, 27, 64, 81, 92

speech, 39

speed of light, 61, 64, 66

spin, vii, 14, 17, 21, 26, 29, 35, 40, 49, 78, 79, 113, 120, 141, 142, 143

spindle, 104

spintronic devices, 143

stability, viii, 1, 2, 3, 6, 7, 9, 11, 19, 29, 32, 54, 79, 82, 104, 127, 134

stabilization, 78, 79

stabilizers, 44, 123
standard error, 82
state(s), 17, 19, 22, 23, 24, 26, 39, 40, 42, 46, 47, 48, 53, 66, 67, 69, 73, 75, 84, 87, 120, 140, 143
statistics, 46
steel, 98
storage, vii, 15, 41, 60, 120, 135
storage media, vii, 41
storms, 103
stress, 129
stretching, 84
stroke, 101
structure, vii, 29, 40, 42, 47, 67, 81, 82, 110, 142
structuring, 127
styrene, 34
substitution, 49
substrate(s), 26, 43, 86, 124, 131
sulfate, 128
Sun, 95, 137
supermolecular structures, 131
superparamagnetic, 21, 23, 26, 27, 42, 45, 77, 79, 86, 88, 93, 99, 130
suppression, 92, 127
surface area, vii, 40, 41, 78, 134
surface chemistry, 128
surface modification, 44, 78, 79, 91
surface properties, 27, 123
surface treatment, 79
surfactant(s), 2, 3, 6, 7, 8, 27, 30, 32, 33, 34, 35, 53, 75, 78, 79, 83, 87, 93, 126, 127, 128, 144, 145
susceptibility, 16, 17, 26, 55, 56, 61, 62, 63, 64, 99
suspensions, 53, 54, 69, 71, 73, 74, 78, 108, 121
symmetry, 22, 23, 24, 55, 58, 63, 69, 139, 140, 141
synthesis, vii, viii, ix, 1, 8, 29, 30, 32, 33, 34, 35, 39, 41, 43, 44, 50, 51, 77, 78, 80, 91, 104, 108, 119, 121, 123, 125, 127, 128, 129, 130, 131, 135, 139, 143, 144, 145

T

targeted drug delivery, vii, ix, 9, 78, 83, 97, 100, 135
techniques, viii, ix, 15, 29, 33, 35, 43, 55, 56, 78, 79, 124, 130, 143, 144, 145
technological revolution, 120
technology(s), vii, 2, 40, 41, 45, 60, 98, 108, 128
TEM, 81, 83, 84, 85, 89, 93, 123, 127, 128, 129, 132, 133, 134
temperature, 1, 9, 11, 15, 21, 22, 23, 24, 25, 26, 30, 34, 36, 42, 43, 44, 69, 77, 79, 80, 81, 86, 87, 88, 90, 91, 98, 99, 100, 104, 135, 140, 142
TEOS, 80, 81
testing, 92
therapy, 88, 120, 134

thermal activation, 23, 87
thermal decomposition, 123
thermal energy, 4, 6, 24, 26, 30, 45
thermal stability, 54, 75, 99
thermodynamic equilibrium, 78
thermodynamics, 43
thermolysis, 34
thin films, 26, 46, 125, 140, 143
thinning, 126, 127
threats, 107
throws, 108
tin, 14
tissue, 78, 79, 82, 89
toluene, 32, 34
topology, 26
total energy, 18, 19, 23, 25, 26
toxicity, 79, 91
TPA, 55, 66, 67
trade, 49
transducer, 100
transition metal, 141
translation, 60, 69
transmission, 60, 64, 65, 68, 69, 71, 74, 75, 77, 81
transmission electron microscopy, 77, 81
transparency, 35, 41, 49, 101
transport, 100, 120, 143
treatment, 34, 78, 79, 80, 100, 128, 134
tumor(s), 78
tumor cells, 78
tunneling, vii, 40, 120

U

ultrasound, 34, 80, 123
uniform, 6, 22, 65, 67, 68, 80, 86, 121, 128
universality, 59
USA, 119

V

vaccine, 92
vacuum, 8, 27, 46, 66, 98
valence, 46, 67, 140, 141
valve, 40
vapor, 43
variations, 58
vector, 17, 18, 60, 61, 141
velocity, 5, 9, 64, 110, 113
versatility, 43
vibration, 80
viruses, 125
viscosity, 6, 7, 24, 79, 81, 89, 90, 98-100, 126, 129

Index 157

vision, 65
volatility, 99

W

waste management, 127
water, 2, 3, 7, 8, 30, 32, 33, 34, 35, 77, 78, 79, 80, 81, 90, 93, 98, 99, 100, 121, 135, 144, 145
water purification, 135
wavelengths, 69, 71, 100, 101
wear, 98, 99
wells, 44, 136
wetting, 130, 131, 135
wires, 42, 47, 136
writing process, 9

X

X-ray diffraction (XRD), 68, 81, 82, 83, 84, 89, 93, 135

Y

yield, 43, 67

Z

zinc, 8, 113, 117, 145
zinc oxide, 145